Sitzungsberichte der Heidelberger Akademie der Wissenschaften

Mathematisch-naturwissenschaftliche Klasse

Die Jahrgänge bis 1921 einschließlich erschienen im Verlag von Carl Winter, Universitäts-buchhandlung in Heidelberg, die Jahrgänge 1922—1933 im Verlag Walter de Gruyter & Co. in Berlin, die Jahrgänge 1934—1944 bei der Weiß'schen Universitätsbuchhandlung in Heidelberg. 1945, 1946 und 1947 sind keine Sitzungsberichte erschienen.

Ab Jahrgang 1948 erscheinen die „Sitzungsberichte" im Springer-Verlag

Inhalt des Jahrgangs 1948:

1. P. Christian und R. Haas. Über ein Farbenphänomen. DM 1.50.
2. W. Blaschke. Zur Bewegungsgeometrie auf der Kugel. DM 1.—.
3. P. Uhlenhuth. Entwicklung und Ergebnisse der Chemotherapie. DM 2.—.
4. P. Christian. Die Willkürbewegung im Umgang mit beweglichen Mechanismen. DM 1.50.
5. W. Bothe. Der Streufehler bei der Ausmessung von Nebelkammerbahnen im Magnetfeld. DM 1.—.
6. W. Troll. Urbild und Ursache in der Biologie. DM 1.50.
7. H. Wendt. Die Jansen-Rayleighsche Näherung zur Berechnung von Unterschall-strömungen. DM 2.40.
8. K. H. Schubert. Über die Entwicklung zulässiger Funktionen nach den Eigen-funktionen bei definiten, selbstadjungierten Eigenwertaufgaben. DM 1.80.
9. W. Schaaff. Biegung mit Erhaltung konjugierter Systeme. DM 1.80.
10. A. Seybold und H. Mehner. Über den Gehalt von Vitamin C in Pflanzen. DM 9.60.

Inhalt des Jahrgangs 1949:

1. H. Maass. Automorphe Funktionen und indefinite quadratische Formen. DM 3.60.
2. O. H. Erdmannsdörffer. Über Fasergranite und Böllsteiner Gneis. DM 1.20.
3. K. H. Schubert. Die eindeutige Zerlegbarkeit eines Knotens in Primknoten. DM 2.80.
4. K. Holldack. Grenzen der Herzauskultation. DM 4.20.
5. K. Freudenberg. Die Bildung ligninähnlicher Stoffe unter physiologischen Be-dingungen. DM 1.—.
6. W. Troll und H. Weber. Morphologische und anatomische Studien an höheren Pflanzen. DM 7.80.
7. W. Doerr. Pathologische Anatomie der Glykolvergiftung und des Alloxandiabetes. MD 9.80.
8. W. Threlfall. Knotengruppe und Homologieinvarianten. DM 1.50.
9. F. Oehlkers. Mutationsauslösung durch Chemikalien. DM 3.80.
10. E. Sperner. Beziehungen zwischen geometrischer und algebraischer Anordnung. DM 3.—.
11. F. Heller. Ursus (Plionarctos) stehlini Kretzoi. DM 4.80.
12. W. Rauh. Klimatologie und Vegetationsverhältnisse der Athos-Halbinsel und der ostägäischen Inseln Lemnos, Evstratios, Mytiline und Chios. DM 10.50.
13. Y. Reenpää. Die Schwellenregeln in der Sinnesphysiologie und das psychophysische Problem. DM 1.60.

Sitzungsberichte
der Heidelberger Akademie der Wissenschaften
Mathematisch-naturwissenschaftliche Klasse
Jahrgang 1962/63, 2. Abhandlung

Symposium
über Automation und Digitalisierung in der
Astronomischen Meßtechnik

am 27. und 28. April 1962 in Tübingen

Mit 51 Textabbildungen

Im Auftrag der Heidelberger Akademie der Wissenschaften
herausgegeben von

H. Siedentopf

1963
Springer-Verlag Berlin Heidelberg GmbH

ISBN 978-3-662-22449-6   ISBN 978-3-662-22448-9 (eBook)
DOI 10.1007/978-3-662-22448-9

Symposium über Automation und Digitalisierung in der astronomischen Meßtechnik

am 27. und 28. April 1962 in Tübingen

Mit 51 Textabbildungen

Im Auftrag der Heidelberger Akademie der Wissenschaften

herausgegeben von

H. Siedentopf

Inhaltsübersicht

H. Siedentopf (Tübingen): Automation und Digitalisierung in der astronomischen Meßtechnik. (Mit 4 Textabbildungen.)

Zusammenfassung: Ausgehend von der informationstheoretischen Situation bei astronomischen Beobachtungen wird betrachtet, bei welchen Stellen des Empfangs und der Verarbeitung von Informationen aus dem Weltall, die von Lichtquanten bzw. elektromagnetischen Wellen getragen werden, eine Automatik und die Einführung digitaler an Stelle von analoger Ausgabe von Meßdaten zweckmäßig ist. Von wesentlicher Bedeutung für diese Frage ist die Zahl der in einem Meßprogramm zu verarbeitenden Informationseinheiten.

Als Beispiele werden verschiedene Methoden der lichtelektrischen Photometrie betrachtet, die im Tübinger Astronomischen Institut in Benutzung sind.

Die von Lichtquanten oder elektromagnetischen Wellen aus dem Weltraum kommenden Informationen lassen sich beschreiben durch eine vom Ort an der Sphäre, von der Wellenlänge und der Zeit abhängige Intensität für eine bestimmte Polarisationsrichtung $I_1(x, y, \lambda, t)$. Beim Durchgang durch die Atmosphäre wird diese Intensitätsfunktion geändert und ihr Informationsgehalt verkleinert. Zu diesen Verlusten tragen bei die orts-, wellenlängen- und zeitabhängige Schwächung $k(x, y, \lambda, t)$, die durch die turbulente Struktur der Atmosphäre bewirkte Intensitätsmodulation, die Richtungsänderungen (Refraktion und Richtungsszintillation) und die Strahlungsemission der Atmosphäre (Airglow, Streulicht, thermische Strahlung). Eine Empfangsanordnung oberhalb der Atmosphäre (Satellitenteleskop) ist von diesen Störeffekten frei, unterliegt aber anderen Störungen, z. B. von der kosmischen Partikelstrahlung, und sie wirft neue Probleme der Datenübertragung und Richtungskontrolle auf.

Die Empfangsanordnung gibt einen weiteren Informationsverlust, der von Absorptionsverlusten, von der spektralen Empfindlichkeit und vom Eigenrauschen der Empfänger und von dem begrenzten Auflösungsvermögen für Richtung, Wellenlänge und Zeit herrührt. Die Auflösungsgrenzen lassen sich in erster Näherung durch Parameter Δx, Δy, $\Delta \lambda$, Δt mit der Bedeutung von Halbwertsbreiten bzw. Zeitkonstanten beschreiben. Zu einer genaueren Charakterisierung ist der aus der Nachrichtentechnik geläufige Begriff der Übertragungsfunktion zweckmäßig; sie gibt den bei einer Fourierentwicklung der einfallenden Intensität nach Ort, Wellenlänge und Zeit auftretenden Amplitudenverlust als Funktion

der Fourierfrequenz. Die Verhältnisse werden weiter dadurch
kompliziert, daß die Übertragungsfunktionen im allgemeinen auch
von der Intensität abhängen.

Die Empfangsanordnung hat neben der Umwandlung der einfallenden Intensität in meßbare Größen noch weitere Funktionen
zu erfüllen. Sie trifft aus der Fülle der angebotenen Informationen

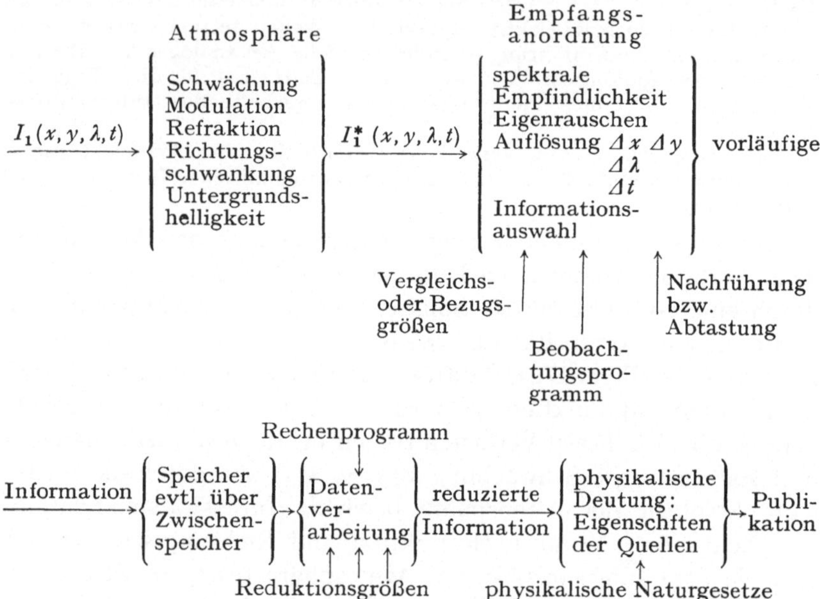

Abb. 1. Schema des Informationsflusses bei astronomischen Beobachtungs- und Reduk-
tionsverfahren

eine Auswahl, je nachdem welches Ziel mit der Beobachtung ver-
folgt wird, z. B. ob es sich lediglich um eine Positionsbestimmung
oder nur um Intensitätsmessung in einem bestimmten Wellen-
längenintervall handelt. Der Empfangsanordnung werden bestimmte Vergleichs- oder Bezugsgrößen zugeführt, diese können
kosmischer oder irdischer Art sein: z. B. die Helligkeiten von Ver-
gleichssternen bei der Bestimmung der Lichtkurve eines Veränderlichen oder die Zenit-Nadir-Richtung bei der Deklinationsmessung
am Meridiankreis.

Da die zu messenden Himmelskörper sich relativ zur optischen
Achse der Empfangsanordnung bewegen, ist eine Nachführungs-
vorrichtung erforderlich, die die optische Achse auf eine bestimmte Richtung des Himmelsgewölbes festhält. Die Nach-

führung ist relativ einfach, wenn die Empfangsanordnung mit dem Erdkörper verbunden ist, sie wird sehr kompliziert bei Messungen von hochfliegenden Ballonen, Raketen und Satelliten aus. In manchen Fällen kann die Nachführung ersetzt werden durch eine ein- oder zweidimensionale Abtastung relativ zu einer bestimmten Bezugsrichtung. Diese Verfahren sind vor allem in der Radioastronomie von Bedeutung geworden, sie spielen aber auch in der optischen Flächenphotometrie eine Rolle.

Der Empfangsanordnung muß ein Meßprogramm vorgegeben werden, das die Einstellungen und die Meßaufgaben festlegt. Hier bietet sich ebenso wie bei der Nachführung die erste Möglichkeit einer Automatisierung. Als Beispiel erwähne ich eine ganz einfache Automatik bei unseren zur Zeit in Südafrika laufenden Messungen der Polarisation, Helligkeit und Farbe im Zodiakallicht durch lichtelektrische Photometrie mit einem azimutal aufgestellten Doppelfernrohr. Es wird in Schnitten parallel zum Horizont gemessen. Der elektromechanische Programmgeber stellt ein vorgegebenes Azimut ein, läßt in dem einen optischen System 12 sec lang die Polarisation bei rotierendem Polarisator registrieren, im anderen System je 6 sec lang die Helligkeit mit Gelb- und mit Blaufilter. Dann erfolgt der Übergang zum nächsten Azimut, und während dieses Übergangs werden für beide Systeme die Dunkelströme der Multiplier aufgezeichnet. Die Aufzeichnung erfolgt mit Potentiometerschreibern, also analog; wir haben es mit einer automatischen, aber nicht digitalen Meßvorrichtung zu tun.

Die von der Empfangsanordnung ausgewählte und gemessene Information wird in jedem Fall einem Speicher zugeführt. Als Speicher dient bei direkter Ablesung und Aufschreibung der Meßwerte das Beobachtungsbuch, bei objektiver Aufzeichnung können die Informationen analog in Form von Registrierkurven, digital in Form von ausgedruckten Ziffern, Lochkarten oder Lochstreifen gespeichert werden. Auch das Magnetband läßt sich als Speicher — sowohl analog wie digital — verwenden. Der wichtigste Speicher ist in der Astronomie immer noch die photographische Schicht. Das liegt an ihrer großen Informationskapazität, die rund 6×10^5 bits pro cm² oder rund 3×10^8 bits auf einer 24×24 cm²-Platte beträgt. Bei Sternaufnahmen wird diese Kapazität allerdings nicht ausgenutzt, da nur ein kleiner Teil der Speicherplätze mit Bildern von Sternen besetzt ist; dagegen wird bei Photographien der Sonnengranulation oder anderer flächenhafter Objekte die Infor-

mationskapazität der Schicht voll in Anspruch genommen. Eine
gewisse Konkurrenz erwächst der photographischen Schicht neuer-
dings im Superorthikon mit verlängerter Speicherzeit, bei dem sich
auf der Speicherplatte Informationen bei Summationszeiten von
100 oder mehr Sekunden für mehrere Stunden speichern lassen.
Wegen der besseren Quantenausbeute und der Möglichkeit der
Vorverstärkung durch einen elektronischen Bildverstärker läßt
sich etwa die 100fache Empfindlichkeit gegenüber dem photographischen Prozeß erzielen, wobei das Auflösungsvermögen fast
gleichwertig zur Photographie, die Speicherfläche dagegen wesentlich kleiner, von der Größenordnung wenige cm^2 ist. Die Abnahme
der Information von der Speicherplatte muß durch zeilenweise
Abtastung wie beim Fernsehen erfolgen, ein ähnliches Abtast-
verfahren mit Hilfe eines Lichtpunktlinienabtasters läßt sich auch
zur automatischen Auswertung von Photoplatten verwenden. Als
Speichersystem erhöhter Empfindlichkeit muß in diesem Zusammenhang noch die elektronische Kamera von Lallemand genannt
werden, die bei ausgezeichnetem Auflösungsvermögen eine etwa
30fache Empfindlichkeitssteigerung gegenüber einer höchstempfind-
lichen photographischen Schicht erreicht. Wie beim Superorthikon
bedeuten die kleine Speicherfläche und der beträchtliche Aufwand
Nachteile gegenüber der direkten Photographie.

Mit der Aufnahme der von der Empfangsanordnung gelieferten
vorläufigen Informationen in den Speicher ist die Beobachtungs-
aufgabe noch nicht beendet. Der nächste Schritt ist die Datenverarbeitung, die den Zweck hat, die Einflüsse der Atmosphäre
und der Empfangsanordnung soweit als möglich zu eliminieren und
die gewünschten Informationsdaten in sauber reduzierte Form zu
bringen. Die Datenverarbeitung wird sehr erleichtert, wenn die
vorläufigen Informationen dem Speicher in digitaler Form, sei es
als Lochkarte oder als Lochstreifen entnommen werden können.
In diesem Fall kann die Verarbeitung mit Hilfe einer digitalen
elektronischen Rechenanlage erfolgen, in die neben den Speicher-
daten die nötigen Reduktionsgrößen und ein Rechenprogramm
eingegeben werden.

Ein besonderer Fall liegt vor, wenn eine photographische
Schicht als Speicher verwendet wurde. Die Schicht dient dabei als
Zwischenspeicher, aus der die gesuchten Informationen über Position und Helligkeit durch besondere Meßgeräte entnommen und
vor der Datenverarbeitung einem weiteren Speicher zugeführt

werden. Auch hierbei erleichtert es die Auswertung, wenn die Ausgabe, sei es bei einem Irisblendenphotometer, einem Koordinatenmeßgerät oder einem Registrierphotometer zur Auswertung von Spektren in digitaler Form erfolgt.

Eine andere Art von Zwischenspeicher bildet das Magnetband, wie es z. B. bei Messungen aus Satelliten zur vorläufigen Speicherung der Meßdaten bis zum Abruf durch eine Bodenstation dient. Die Speicherkapazität von Magnetband ist beträchtlich, man braucht nur daran zu denken, daß eine Fernsehsendung mit einem Informationsfluß von einigen Millionen bits/sec sich auf Magnetband übertragen oder von Magnetband abspielen läßt. Auch für die datenverarbeitenden Maschinen hat das Magnetband immer mehr an Bedeutung gewonnen. Für eine schnelle Datenausgabe, etwa von einem Digitalvoltmeter mit mehr als 10^3 bits/sec ist das Magnetband unentbehrlich.

An sich ist jeder Zwischenspeicher von Nachteil, weil bei der Speicherung und nachfolgenden Datenübertragung immer ein gewisser Informationsverlust eintritt. Dieser Verlust wiegt aber gegenüber dem großen Vorteil, den vor allem die photographische Schicht für die astronomische Beobachtungstechnik bietet, nicht sonderlich schwer.

Mit der reduzierten Information ist der eigentliche Beobachtungsprozeß abgeschlossen, und manche Publikationen, z. B. der Lichtkurve eines Veränderlichen oder der Positionen eines Kometen enthalten nur solche reduzierten Informationen. Das nächste Glied in der Kette, die zur Erkenntnis astronomischer Tatbestände führt, ist die physikalische Deutung der empfangenen Informationen. Uns interessieren ja nicht die Positionen eines Kometen, sondern : eine Bahnkurve, nicht die Lichtkurve eines Bedeckungsveränderlichen, sondern die Eigenschaften seiner Komponenten. Die von den elektromagnetischen Wellen aus dem Kosmos gebrachten Informationen enthalten die gesuchten Eigenschaften der Himmelskörper in verschlüsselter, kodierter Form. Um sie zu entschlüsseln, um aus ihnen auf die Eigenschaften der Quellen schließen zu können, müssen wir die physikalischen Naturgesetze anwenden, deren universelle Gültigkeit in Raum und Zeit die entscheidende Voraussetzung der astronomischen Forschung ist.

Im allgemeinen braucht man eine größere Anzahl reduzierter Informationen, um zu einer physikalischen Deutung der Beobachtungsbefunde zu gelangen; von den drei Kometenpositionen zur

vorläufigen Bahnbestimmung bis zu den vielen tausenden von
Konturen der 21 cm-Wasserstofflinie zur Bestimmung der Verteilung des neutralen Wasserstoffs in einem Bereich des Milch-
straßensystems gibt es alle Zwischenstufen.

Wir wollen nun fragen, an welchen Stellen der Kette von der
Intensitätsaufnahme bis zur physikalischen Deutung eine Automatisierung angezeigt ist und unter welchen Bedingungen der
Aufwand dafür lohnt. Wie erwähnt liegt die erste Möglichkeit bei
der Automatisierung der Empfangsanordnung, sei es durch automatische Nachführung oder durch Programmsteuerungen für die
Einstellung der Beobachtungsobjekte. Der zweite Schritt ist die
Ausgabe und die Speicherung der vorläufigen Informationen in
digitaler Form, damit die Reduktion ohne weitere Umsetzung mit
digitalen programmgesteuerten Datenverarbeitungsanlagen erfolgen
kann. Bei Verwendung eines Zwischenspeichers ist die automatische
und digitale Ablesung der Informationen aus dem Zwischenspeicher
— photographische Schicht, Magnetband — anzustreben.

Wie weit man bei einem Meßvorhaben mit der Automatisierung
gehen soll, hängt vom Umfang der gestellten Aufgabe ab. Es ist
allerdings schwer, eine Grenze anzugeben, von der ab die Digitalisierung der Messungen und die automatische Auswertung lohnen.
Die Grenze hängt nicht so sehr von der Zahl der zu messenden
Objekte ab als von dem Umfang der zu verarbeitenden Informationseinheiten. Als eine grobe Abschätzung kann vielleicht
gelten, daß bei mehr als 10^5 bits Informationsgewinn die digitale
Messung mit programmgesteuerter Datenverarbeitung immer
zweckmäßig ist, während bei weniger Informationseinheiten der
Aufwand nicht lohnt. Als Beispiel sei angeführt die photographische
Photometrie eines offenen Sternhaufens von 100 Mitgliedern in
drei Farben, wobei vier Platten für jede Farbe aufgenommen
werden. In diesem Fall sind rund 12000 bits Informationseinheiten
zu verarbeiten; für die Auswertung eines einzelnen Haufens, sozu-
sagen als Gelegenheitsarbeit, würde sich der Ausbau eines Iris-
photometers zu einer automatischen Anlage mit digitaler Ausgabe
nicht lohnen. Dagegen sollte ein Institut, bei dem die photographische Sternphotometrie zum täglichen Handwerk gehört und
das jeden Tag einige tausend bits dabei verarbeitet, unbedingt die
Automatisierung der Photometrie anstreben.

Der nicht unbeträchtliche Aufwand für den Bau automatischer
digitaler Meß- und Auswertgeräte wird durch den Zeitgewinn und

die Einsparung von Arbeitskräften im allgemeinen wettgemacht. Dagegen ist mit der Automation ein weiteres Problem verbunden, das einer sorgfältigen Diskussion bedarf: die Frage der Zuverlässigkeit und der notwendigen Kontrollen. Wenn bei der lichtelektrischen Photometrie von Sternen die Spannung am Arbeitswiderstand des Multipliers mit einem Potentiometerschreiber registriert wird, so erkennt man aus der Registrierkurve sofort, ob bei den Messungen instrumentell und atmosphärisch alles in Ordnung war. Wenn ich dagegen die über die Meßdauer gemittelte Spannung über geeignete Umsetzer auf Lochstreifen ausdrucke, so sehe ich nichts mehr. Es wird daher erforderlich, eine genügende Zahl von Kontrollen einzubauen, um sicher zu sein, daß instrumentell keine Störung aufgetreten ist. Ungewöhnliche Helligkeitsschwankungen durch Dunststreifen oder Cirren müssen besonders angezeigt werden. Ein Hilfsmittel, das bei dem genannten Beispiel und manchen anderen Meßaufgaben benutzt werden kann, ist die gleichzeitige analoge Registrierung bei digitaler Datenausgabe. Die Registrierung dient lediglich zur Kontrolle, die Digitaldaten zur weiteren Reduktion mit Hilfe einer elektronischen Rechenanlage.

Ich möchte diese allgemeinen Betrachtungen hier abbrechen, bei den Diskussionen wird vielleicht Gelegenheit sein, darauf zurückzukommen. Zum Abschluß meines Berichts will ich noch einige Beispiele von photometrischen Geräten und Meßverfahren erwähnen, die in den letzten Jahren im Tübinger Astronomischen Institut entwickelt und benutzt worden sind, bzw. sich noch in der Erprobung befinden. Von der Industrie wird in letzter Zeit eine wachsende Anzahl von Digitalvoltmetern, Spannungs-Frequenz-Wandlern, elektronischen Zählern mit Umsetzern zur Ausgabe auf Zifferndrucker, Lochkarten und Lochstreifen angeboten, so daß für die meisten Meßaufgaben geeignete Industrie-Geräte vorhanden sind.

Die im Multiplier M erzeugten Elektronenlawinen (Abb. 2) werden durch den Kathodenfolger KF in negative Impulse verwandelt, durch einen Linearverstärker LV auf maximal 50 V verstärkt und einem Diskriminator D zugeführt, der nur die Impulse über etwa 8 V Höhe an den 1 MHz-Zähler Z weitergibt. Auf diese Weise wird das Verhältnis von Signal- zu Dunkelstrom etwas verbessert, da die Verteilungsfunktion der Dunkelimpulse ihr Maximum bei kleinerer Impulshöhe hat als die Verteilungsfunktion der Signalimpulse. Die Torzeiten Δt, über die die einfallende Intensität $I(t)$ integriert wird, lassen sich am Zähler Z (Beckman-Berkeley

Model 7360) einstellen. Die Zählwerte werden mit Hilfe eines Umsetzers U und eines elektromagnetischen Aufsatzes A durch eine Saldiermaschine S ausgedruckt. Diese unter Verwendung kommerzieller Einzelgeräte aufgebaute Anordnung wurde u. a. dazu benutzt, die langsamen Modulationsschwankungen der Stern-

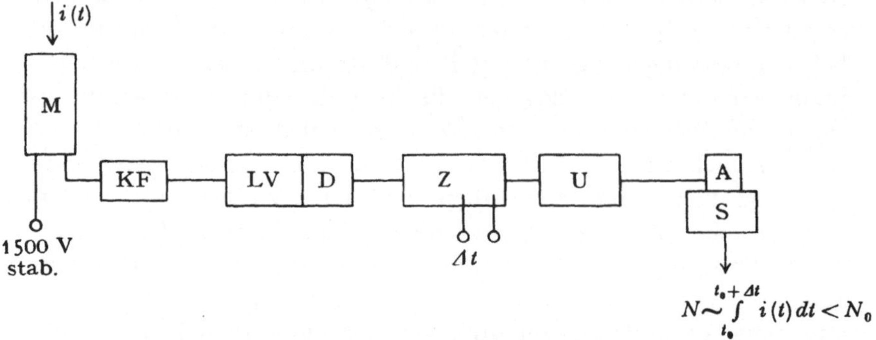

Abb. 2. Blockschaltbild eines lichtquantenzählenden Photometers

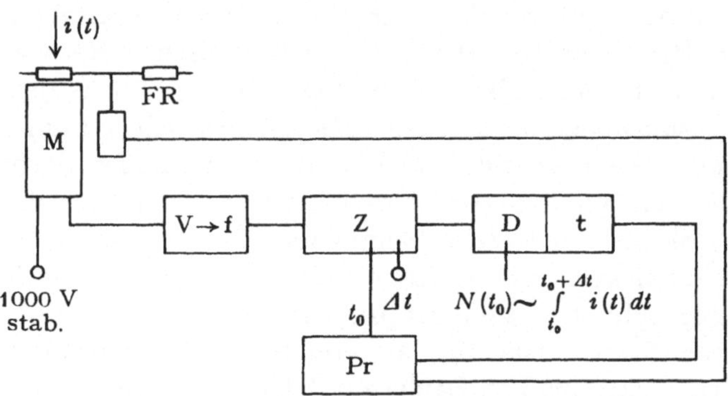

Abb. 3. Blockschaltbild eines automatisch arbeitenden lichtelektrischen Photometers mit digitaler Datenausgabe

helligkeiten zu untersuchen, die die Genauigkeit der lichtelektrischen Sternphotometrie begrenzen.

Das in Abb. 3 skizzierte Photometer dient zur automatischen Aufzeichnung der Leuchtdichte des Nachthimmels im Zenit bei verschiedenen, durch den mit Interferenzfiltern bestückten Filterrevolver FR ausgewählten Wellenlängen. Die am Arbeitswiderstand des Multipliers M entstehende Spannung liegt am Eingang eines Spannungs-Frequenzwandlers Dymec Model 2210, der die Spannung 0—1 V in eine ihr proportionale Frequenz von 0—10 kHz umsetzt. Diese Frequenz wird in dem Zähler Z (Hewlett-Packard 521 a) bei fester Torzeit Δt gezählt und mit dem Drucker (Hewlett-Packard 560 A) dezimal ausgedruckt. Gleichzeitig wird die von der Digitaluhr U (Hewlett-Packard 570 A) abgegebene Uhrzeit t_0 der Beobachtung gedruckt. Ein Programmgeber Pr besorgt den Filterwechsel und den Start der Messungen. Es werden pro Minute fünf Filterwerte und der Dunkelstrom bei Integrationszeiten von je 10 sec gedruckt. Ein Analogausgang des Druckers ermöglicht zur Kontrolle die gleichzeitige Analogregistrierung der Meßwerte auf

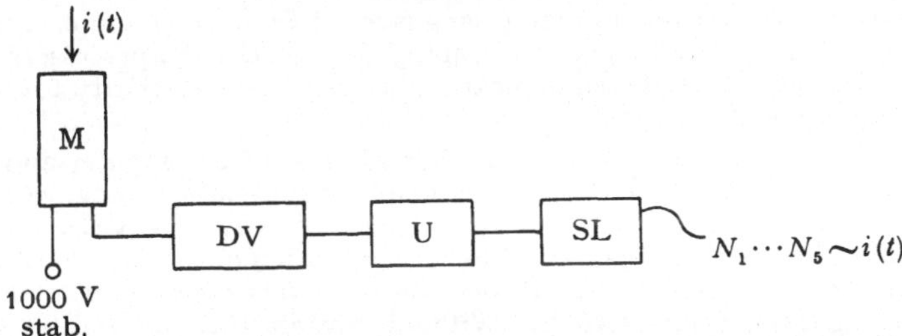

Abb. 4. Blockschaltbild eines Photometers mit Digitalvoltmeter und Lochstreifenausgabe

einem Linienschreiber. Die Anordnung arbeitet nach dem Einschalten vollautomatisch und liefert ein vollständiges Meßprotokoll.

Ein sehr einfaches lichtelektrisches Photometer (Abb. 4) erhält man, wenn die Spannung am Multiplierausgang mit Hilfe eines Digitalvoltmeters DV gemessen und der Meßwert über einen Umsetzer U auf Lochstreifen oder Lochkarten gestanzt wird. Die bei uns erprobte Anordnung verwendet ein Beckman Digitalvoltmeter Model 5350 H, einen Umsetzer Model G 3102 Z der gleichen Firma und einen Schnellocher Typ SL 614 von Standard Elektrik Lorenz. Die Anordnung gestattet 4—5 Messungen pro Sekunde; entgegen den in Abb. 2 und 3 gezeigten Photometern ist aber keine zeitliche Integration möglich, sondern nur die Messung der Momentanwerte des Photostroms. Der Photostrom muß also hinreichend groß und nicht durch Rausch- und Szintillationsschwankungen beeinflußt sein. Als Anwendung kommt z. B. die Schwärzungsmessung bei photographischer Photometrie in Betracht. Natürlich lassen sich die Geräte U bis SL auch mit den integrierenden Photometern der Abb. 2 und 3 kombinieren. Ihr besonderer Vorteil liegt darin, daß die auf Lochstreifen ausgegebenen Meßwerte sogleich einer datenverarbeitenden Anlage zur weiteren Reduktion zugeführt werden können.

H. G. Walter (Munich): The Importance of the Programming Language ALGOL for Automation in Astronomical Research.

Summary: For the evaluation of astronomical observations electronic computers are more and more in use. Generally, however, these computers differ in their inherent code of programming. Therefore, a program written in this code is restricted to a special computer. If on the contrary an algorithm is coded in the programming language ALGOL, a program of such a kind is universal for all computers equipped with ALGOL-processors and it acts as a general tool of communication because of its similarity to customary mathematical notation.

The connecting link between the inherent code of a computer and the programming language ALGOL is the principle of automatic coding, which is described in the present paper. Further it gives a survey of ALGOL. — The advantages of ALGOL are explained by an application of the theory of MIE. The computations based on this theory can be standardized by writing the required algorithms in ALGOL. Without any further expenditures of programming such calculation methods can be employed by the numerous research-places working on light scattering.

1. Introduction

The work in astronomy generally consists of two sections: Firstly, an astronomer has to perform the observations on the subject to be investigated and to record the information received from them. To some extent this process can be done automatically. Secondly, he has to evaluate the recorded information by means of mathematical methods. Up to a couple of years ago this work had to be accomplished manually and therefore became very troublesome. Nowadays, however, the electronic digital computers are so widely spread that the intrinsic job of evaluation of an immense quantity of measurements and the calculation of huge tables is greatly facilitated.

In order to communicate with a computer one has to write a program. It is a great progress that writing of a program can be performed automatically. This principle of programming is called automatic coding. It is for this reason that we can say that also the second section of the work — the evaluation part — can be settled automatically, provided that the mathematical method is given. In this connexion ALGOL has something to do with automation in astronomy.

2. The Principle of Automatic Coding

Considering the working-method of a digital computer one realizes that normally an input string of characters is transformed by a machine-program into an output string of characters. From this perception issues the usage of a computer for translation purposes.

In principle the following situation is present: An electronic digital computer is instructed in a certain logical language, i.e. the machine language or the internal code of the machine. A program written in this internal code must be stored onto the memory of the particular computer so that it is able to carry out the computation without any help from outside. However, the internal code is not well suited for human beings, because it differs very much from customary notations, and it causes programming to be sometimes an onerous job. To remove this bottleneck of programming one makes use of the fact, that a computer is able to perform all logical operations and consequently is also able to translate one language into another, especially to translate a non-machine language into the machine language. The latter case is fundamental for what is called automatic coding and will be explained below.

For scientific purposes one needs a so called problem-oriented language in which mathematical processes can be described conveniently and which above all is machine-independent. An example of such a language is the programming language ALGOL (Algorithmic language).

If a computer has to carry out computations described in ALGOL, it has to accomplish two tasks. On the one hand, it acts as a translator controlled by the translating program, the input of which is the actual program written in ALGOL and the output of which is the translation of the ALGOL-program into the machine language. The translation — the generated program — needs not to be printed out; more preferable in most cases is its storage in the computer memory for later use. On the other hand, the computer performs the calculations controlled by the machine program which was generated by the above mentioned translating process. At this second stage the machine does exactly the same as if the running program were hand-coded in the internal code of the computer.

Both stages are illustrated in the following flow diagrams:

First stage

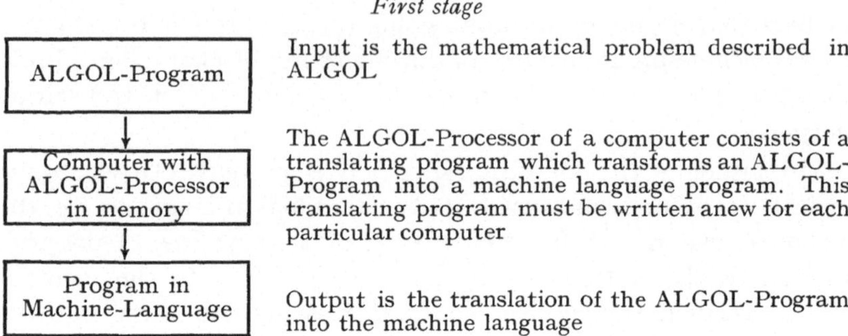

Input is the mathematical problem described in ALGOL

The ALGOL-Processor of a computer consists of a translating program which transforms an ALGOL-Program into a machine language program. This translating program must be written anew for each particular computer

Output is the translation of the ALGOL-Program into the machine language

This process is called automatic coding, since the computer itself codes a program.

Second Stage

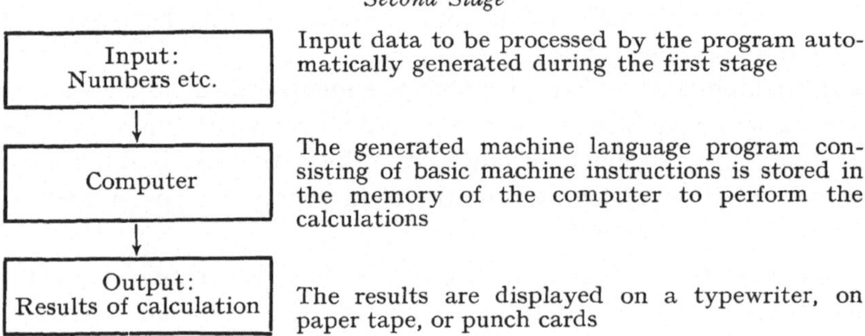

Input data to be processed by the program automatically generated during the first stage

The generated machine language program consisting of basic machine instructions is stored in the memory of the computer to perform the calculations

The results are displayed on a typewriter, on paper tape, or punch cards

3. A Survey of the Programming Language ALGOL

ALGOL has passed through several years of development in which representatives of European countries and the United States of America had participated. Finally it was released in January 1960 and since then it gets more and more in use.

This paper is restricted to the essentials of ALGOL and cannot be more than a brief description of the language and its structure. Excellent publications dealing with the details of ALGOL and giving a systematic introduction to the language are listed under the references [1], [2], [3], [4], [5].

As mentioned in the foregoing ALGOL is a language allowing convenient programming of scientific problems. With this is meant the possibility of direct writing down an ordinary formula, an easy way of stating alternatives and deviations from the normal sequence of execution, facilities for defining loops, etc. The quantities occurring in an ALGOL-program are represented by names as it is customary in arithmetic. A name is a string of letters and digits beginning with a letter. — Statements are the main constituents of an ALGOL-program. Each statement usually corresponds to a number of computer instructions written in the machine language.

The assignment statement

$$y := 1/2 \times (a+b)$$

indicates that the new quantity "y" is to be calculated as the half sum of "a" and "b", where both "a" and "b" are previously introduced. The symbol "$:=$" may be interpreted as an arrow leading from right to left and indicating that the value of the formula on the right-hand side is assigned to the quantity written on the left-hand side of an assignment statement. Each calculation in ALGOL is arranged in this manner.

Some more complicated assignment statements are presented in the complete ALGOL-program of chapter 4.

A conditional statement is proper to express an alternative:

$$\text{if } x > 1 \text{ then } z := 1/x.$$

The statement $z := 1/x$ is executed if $x > 1$; otherwise it is skipped. An example for a real alternative would be

$$\text{if } x > 1 \text{ then } z := 1/x \text{ else } z := 1-x.$$

The sequence of execution of statements may be interrupted by a go to statement. The go to statement

$$\text{go to PP 3}$$

means that the next statement to be performed is the statement with the label PP 3; for instance

$$\text{PP 3:   for } v := 1 \text{ step } 0.5 \text{ until } 10 \text{ do } w := \sin(v).$$

The above statement is convenient for writing loops; it indicates that the statement $w := \sin(v)$ is to be executed 19 times. Another particularity of ALGOL is shown in this statement: In an ALGOL-program the most frequent functions as sin, cos, arctan, ln, exp, etc. may be incorporated without any additional requirements.

Several statements may be combined to a compound statement by placing these statements between "begin" and "end".

All statements are essentially dynamic, because they depend upon a certain order in which they are carried out. In contrast with statements declarations are static units, the purpose of which is to give some information about the whole program.

As an example, the declaration

$$\text{integer } n, v$$

specifies the quantities n and v to be integers and throughout the program they are treated as such.

The definition of subprograms is rendered possible by procedures, which represent a special kind of declaration. A procedure is linked with the main program by means of a procedure statement. A detailed explanation of procedures falls outside the scope of this paper and is given in some of the referenced publications [3], [4], [5].

4. The Advantages of ALGOL

Since ALGOL is a machine-independent, problem-oriented language with general adoption it is an international language and therefore acts as an universal tool of communication. Moreover, an algorithm written in ALGOL is well comprehensible and, beyond it, because ALGOL bridges the gap between computer and classical mathematical notation, each ALGOL-program runs on each computer equipped with an ALGOL-translator. An example may explain this situation:

The calculation of light scattering on spherical particles is executed according to the theory of Mie [6] at numerous places and by means of great variety of digital computers. The number of different programs written for this purpose is equal to the number of various computers. Accordingly, as matters stand a standardization of programming work is desirable and it is ALGOL that profits here. For all these programs on light scattering are replaceable by one common ALGOL-program which, besides, could be written more rapidly than each of the other machine language programs. The program example given below shows the way along which this target can be reached. The notation used in the program example is that of the ALCOR-group[1].

---

[1] ALCOR is an abbreviation for **ALGOL CONVERTER**. The ALCOR-group comprises several institutions which try to achieve agreement in coding of ALGOL-programs [3]. From this complete interchangeability of ALGOL-programs will result.

Example of an ALGOL-Program

'BEGIN' 'COMMENT': COMPUTATION OF MIE-COEFFICIENTS OF REAL REFRACTION INDEX.

AUTHOR: H. G. WALTER, SIEMENS UND HALSKE AG., MUNICH.

INPUT DATA OF THE ALGORITHM ARE THE REFRACTION INDEX M AND THE VALUES OF THE MIE-PARAMETER ALPHA = 2 PI × RADIUS/WAVE-LENGTH, FOR WHICH THE MIE-COEFFICIENTS ARE TO BE CALCULATED.— THE RICCATI-BESSEL FUNCTIONS S AND C ARE COMPUTED RECUR-SIVELY. A RESP. B ATTACHED TO S AND C INDICATES THE ARGUMENT ALPHA RESP. BETA, THE ADDITIONAL WORD 'STRICH' SIGNIFIES THE FIRST DERIVATION. — THE REAL AND IMAGINARY PARTS (RE A, IM A, RE P, IM P) OF THE MIE-COEFFICIENTS A AND P ARE PRINTED OUT TOGETHER WITH THE CORRESPONDING CONTROL INDEX N;

'REAL' M, ALPHA, ALPHA 1, ALPHA 2, BETA,
        SA, SA 1, SA 2, CA, CA 1, CA 2, SB, SB 1, SB 2,
        SA STRICH, CA STRICH, SB STRICH,
        Z 1, Z 2, Z 3, Z 4, FAKTOR,
        RE A, IM A, RE P, IM P;

'INTEGER' V, N;
        READ (M, ALPHA 1, ALPHA 2);

'COMMENT': M, ALPHA 1, AND ALPHA 2 ARE INPUT DATA OF THE ALGO-RITHM. ALPHA 1 IS THE LOWER BOUND AND ALPHA 2 THE UPPER BOUND OF THE ALPHA-INTERVAL FOR WHICH THE COEFFICIENTS ARE TO BE COMPUTED. FOR IDENTIFICATION PURPOSES M, ALPHA 1, AND ALPHA 2 ARE IMMEDIATELY PRINTED;
        PRINT (M, ALPHA 1, ALPHA 2);

'COMMENT': LOOP FOR CALCULATION OF ALL GROUPS OF MIE-COEFFI-CIENTS PERTAINING TO THE ABOVE SPECIFIED ALPHA-INTERVAL;
'FOR' ALPHA : = ALPHA 1 'STEP' 1 'UNTIL' ALPHA 2 'DO'

  'BEGIN'   PRINT (ALPHA, M);

        SA 1 := SIN (ALPHA); CA 1 :=   COS (ALPHA);
        SA 2 := COS (ALPHA); CA 2 := — SIN (ALPHA);

          BETA : = M × ALPHA;

        SB 1 := SIN (BETA); SB 2 := COS (BETA);

          V : = +1;

'COMMENT': CALCULATION OF ONE GROUP OF MIE-COEFFITIENTS
DEPENDING ON A GIVEN VALUE OF ALPHA;

$$N := 1;$$

$$L: \qquad V := -V;$$

$$SA := (2 \times N-1) / ALPHA \times SA\,1 - SA\,2;$$
$$SA\ STRICH := SA\,1 - N / ALPHA \times SA;$$
$$CA := (2 \times N-1) / ALPHA \times CA\,1 - CA\,2;$$
$$CA\ STRICH := CA1 - N / ALPHA \times CA;$$

$$SB := (2 \times N-1) / BETA \times SB\,1 - SB\,2;$$
$$SB\ STRICH := SB\,1 - N / BETA \times SB;$$

$$Z\,1 := ALPHA \times SB\ STRICH \times SA - BETA \quad \times SA\ STRICH \times SB;$$
$$Z\,2 := ALPHA \times SB\ STRICH \times CA - BETA \quad \times CA\ STRICH \times SB;$$
$$Z\,3 := BETA \quad \times SB\ STRICH \times SA - ALPHA \times SA\ STRICH \times SB;$$
$$Z\,4 := BETA \quad \times SB\ STRICH \times CA - ALPHA \times CA\ STRICH \times SB;$$

$$FAKTOR := (2 \times N+1) / (N \times (N+1)) \times V;$$

$$RE\ A := \quad FAKTOR \times Z\,1 \times Z\,2/(Z\,1 \times Z\,1 + Z\,2 \times Z\,2);$$
$$IM\ A := \quad FAKTOR \times Z\,1 \times Z\,1/(Z\,1 \times Z\,1 + Z\,2 \times Z\,2);$$
$$RE\ P := - FAKTOR \times Z\,3 \times Z\,4/(Z\,3 \times Z\,3 + Z\,4 \times Z\,4);$$
$$IM\ P := - FAKTOR \times Z\,3 \times Z\,3/(Z\,3 \times Z\,3 + Z\,4 \times Z\,4);$$

PRINT (N);

PRINT (RE A, IM A);   PRINT (RE P, IM P);

$$SA\,2 := SA\,1; \ SB\,2 := SB\,1; \ SA\,1 := SA; \ SB\,1 := SB;$$

$$CA\,2 := CA\,1; \ CA\,1 := CA;$$

$$N := N + 1;$$

'IF' N 'NOT GREATER' ALPHA + ALPHA/4 'THEN' 'GO TO' L;

'END' ENDING OF ALPHA-LOOP

'END' ENDING OF PROGRAM

The standardization of programming work is one of the facilitations offered by ALGOL: the experts are released from onerous program writing and, nevertheless, they can do it themselves with little expenditure of time by using ALGOL. Moreover, they may interchange the ALGOL-program, since they are easily understandable and readable with little further explanation.

Another facilitation is the flexibility and the ease of maintenance of ALGOL-programs. The above program is illustrating these facts:

one could simply extend it by inserting the formulas for the computation of the efficiency factors for scattering and extinction which are derived from the Mie-Coefficients [7], [8]. In a similar way one could calculate the radiation-pressure or one could generalize the program so that it calculates Mie-Coefficients for particles with complex index of refraction. In all these cases the time to be spent for changing the program is incomparably shorter than the time necessary for maintaining the corresponding machine language program.

References

[1] Backus, J. W. et al.: Report on the algorithmic language ALGOL 60. Numer. Math. 2, 106—136 (1960); — Comm. ACM 3, 299—314 (1960). — [2] Schwarz, H. R.: An introduction to ALGOL. Comm. ACM 5, 82—95 (1962). — [3] ALGOL-Manual der ALCOR-Gruppe. Hrsgeg. von den Mitgliedern der ALCOR-Gruppe, bearbeitet von R. Baumann. Elektronische Rechenanlagen 3, 206—212, 265—269, (1961); 4, 71—85 (1962). — [4] Dijkstra, E. W.: Die Programmierung in ALGOL 60. Elektronische Datenverarbeitung, Beiheft 2, 24—47 (1962). — [5] Bottenbruch, H.: Structure and Use of ALGOL 60. J. ACM 9, 161—221 (1962). — [6] Mie, G.: Ann. Physik 25, 377—445 (1908). — [7] Hulst, H. C. van de: Light Scattering by small particles. New York 1957. — [8] Giese, R.-H.: Die Berechnung der Lichtstreuung an kugelförmigen Teilchen mit einem Digitalrechner. Elektronische Rechenanlagen 6, 240—245 (1961).

Discussion

Fricke: Can you give us some information on development under way to use ALGOL on IBM 7090?

Walter: IBM seems to adhere to its FORTRAN in the first place and is in favour of ALGOL merely in the second place. However, an ALGOL-processor for the IBM 7090 will be implemented. This work is already under way at Urbana University, Urbana, Illinois, USA and is influenced by the ALCOR group. The processor in question will be available for the IBM 7090 machines installed by the Deutsche Forschungsgemeinschaft.

Rohlfs: Would it not be better to keep ALGOL as reference language, but to do the actual computation with a compiler, which is oriented more along the special features of the computer used. Then the programs produced by the compiler could be shorter and faster.

Walter: When ALGOL was developed stress was laid on the reference language as well as on the hardware representation, for one of the targets to be reached by ALGOL was a language which could be translated automatically into machine code. Just this fact facilitates programming and enables the programmer to write programs without taking care of details concerning the special features of a particular computer. Depending on the capacity of the compiler used one has to put up with larger programs and more running time compared with machine coded programs.

Ollongren: In connection with the discussion on the universal applicability of ALGOL it should be remarked that all difficulties with the exchange of ALGOL-programs between various computers originate in the fact that the communication procedures which appear in actual ALGOL programmes are different for different compilers. This fact should not be put to the debt of ALGOL as such since the ALGOL syntax carefully abstains from giving rules as to communication procedures. In actual practice it may be that the number of procedures as mentioned (e.g. reading tape, printing out of results), which are specific for the computer used, and possibly too for the compiler for it, becomes so large that serious difficulties in "adapting" given ALGOL-programs for a certain computer ensues. By then it will become necessary for ALGOL to meet the demands for strictly defined communication procedures.

Walter: At the time being interchanging of ALGOL-programs between different computers requires at least an examination of the input- and output-procedures. However, work on standardizations of these communication procedures is under way, for instance within the ALCOR-group, which is a "hardware"-group with a considerable number of members. By means of these standardized procedures one is enabled to write fairly general output formats entirely in ALGOL.

Høg: Rohlfs said that a serious disadvantage of ALGOL is that it gives slower programs than machine-coded programs. But since most machine time is often spent in very few loops one should think that just rewriting manually these few loops after the compilation would pay, if such a manual change of an ALGOL compiled program is easy.

Walter: Essentially, it must not be that ALGOL-programs are slower than machine coded programs. This depends upon the quality of the compiler employed for the translation process. In all those cases, however, an ALGOL-programmer expects a reduction of the time necessary for the running program he may switch over to machine code in his ALGOL-program by use of "code-procedures". Hence, for including machine code provision is made in ALGOL in virtue of a special element of the language.

Raimond: Agreeing that ALGOL may be of great importance for instance in computing theoretical models, I doubt very much whether ALGOL is suitable for data processing, especially when the input and output consists of millions of numbers.

Walter: The programming language ALGOL is mainly designed for scientific and technical problems in which data processing is not the principal task. By no means ALGOL claims to be approved as a comfortable and convenient data processing language. In so far as a great deal of numbers has to be processed, nevertheless, ALGOL may be used though not too efficiently.

J. Reynen (Brussels): The Ephemeris of Double-Star Relative Radial Velocity Calculated on a Pace 231-R Analogue Computer. (With 5 Figures.)

Summary: The classical equations describing the movements of double stars have been programmed on a PACE 231-R analogue computer.

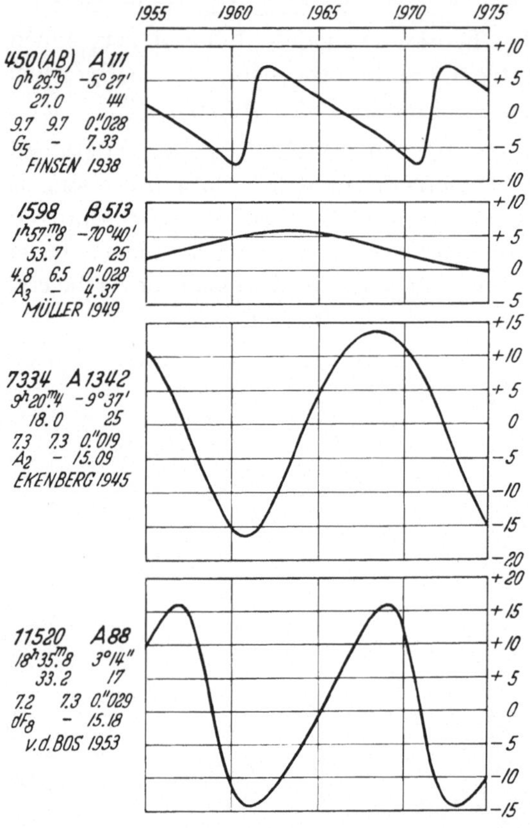

Fig. 1. Relative radial velocities plotted as function of time on special register cards

The output of the computer, being the relative radial velocity of the double star as a function of time, is directly plotted on a special register card. This provides the astronomer an ephemeris of the double star, from which he determines the most favourable time for observation with the spectrometer.

One run, covering a period of 20 years takes about 20 seconds machine time.

The work was initiated by the Observatoire Royal d'Uccle, Brussels, Belgium.

1. Introduction

In order to provide the astronomer with a cataloque of relative radial velocities of the various double stars as function of time, l'Observatoire Royal de Belgique decided to use an analogue computer. In this way the time consuming work of calculating by hand, at certain time intervals, the relative radial velocity could be avoided.

The PACE 231-R analogue computer calculated the relative radial velocity of a double star over one period in about 60 secs. Simultaneously the relative radial velocity is plotted as function of time on a special register card, ready for use in the file (see Fig. 1).

2. *Statement of the Problem*

The movement of double stars is governed by the Kepler equation:

$$E - e \sin E = \frac{2\pi}{P} (t - T) . \tag{1}$$

The relative radial velocity is calculated by:

$$\operatorname{tg} \frac{v}{2} = \sqrt{\frac{1+e}{1-e}} \operatorname{tg} \frac{E}{2} , \tag{2}$$

$$v_r = K \{e \cos \omega + \cos (v + \omega)\} , \qquad K = 29.76 \frac{a'' \sin i}{P p'' \sqrt{1 - e^2}} . \tag{3}$$

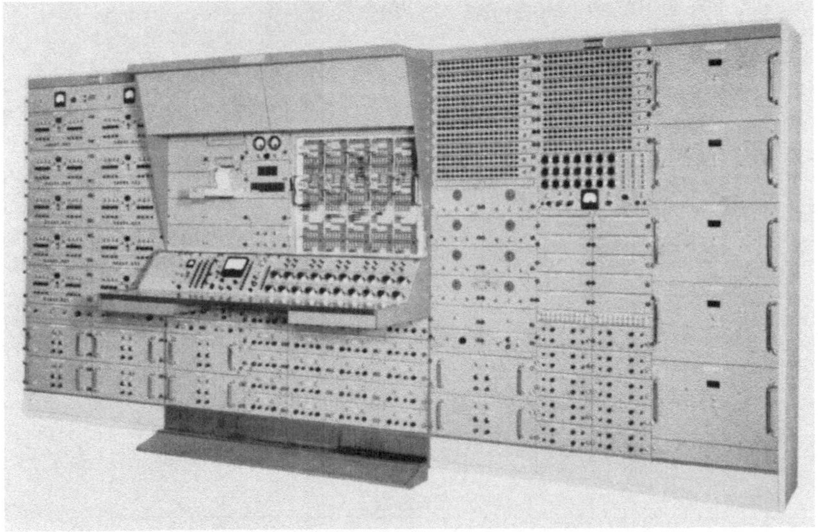

Fig.2. The PACE 231-R analogue computer

The notation is as follows:

E	= eccentric anomaly	i	= inclination
e	= eccentricity	a''	= semi-axis major (sec of arc)
P	= period (years)		
t	= time	p''	= parallax (sec of arc)
T	= time of periastron passage	K	= amplitude (km/sec)
v	= true anomaly	v_r	= relative radial velocity (km/sec).
ω	= the position angle of periastron in true orbit		

3. *Description of PACE 231-R Analogue Computer*

The PACE 231-R analogue computer (Fig. 2) having a basic accuracy of 10^{-4}, reaches this accuracy by means of a temperature controlled oven, taking care that the components governing the accuracy remain at constant temperature. By appropriate connecting of various computation components, an electronic system is

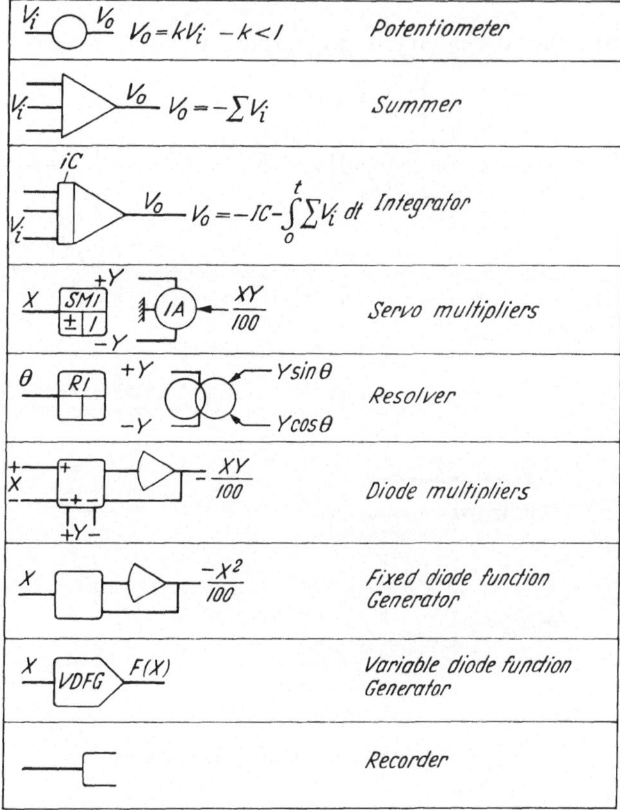

Fig. 3. The various components of the analogue computer

build, governed by the same equations as the physical system, which has to be simulated. The various components are (see Fig. 3)

— potentiometers
— summers
— integrators
— diode multipliers
— servo multipliers
— resolvers (sine and cos winded servo potentiometers)
— fixed function generators for example x^2, x^4, log x, etc.
— variable function generators.

As above mentioned components indicate, problems described by differential equations as well as by ordinary algebraic equations can be simulated. Due to the inherent features of analogue computers, the former are in favour, and as a matter of fact, the possibility of integrating continuously any variable is the most powerful feature of analogue computers. This in contrary with digital computers, where an integration is approached by a numerical process and the accuracy does therefore depend on the integration step used.

4. Description of the Programme

Equation (1) results from an analytical integration with respect to time of Newtons Law, applied on the mechanism of the movement of double stars. As outlined above, the big power of analogue computers is the capability of exact integrating and by doing the integration analytically, this feature of analogue computers is as a matter of fact by-passed.

Indeed, the solution of algebraic equations as (1), (2) and (3) on an analogue computer is usually obtained by converting them to differential equations, of which the steady state solution obeys the algebraic relations to be solved. Differentiating equations (1) with respect to time, results into the equilibrium equation. The analogue computer has become a simulator, simulating the dynamics of the movement of double stars. However, having been able to integrate with respect to any other variable is possible, to obtain a differential equation, which solution obeys the original algebraic relations. For example, differentiating equation (2) with respect to the true anomaly v yields:

$$\frac{dE}{dv} = \frac{\sqrt{1-e^2}}{1+e\cos v} . \tag{4}$$

Further eliminating t and $\sin E$ from (1) yields:

$$t = T + \frac{p}{2}\left(E - \frac{\sqrt{1-e^2}\sin v}{1+e\cos v}\right). \tag{5}$$

From equations (3), (4) and (5), the relative radial velocity can be calculated provided K, ω, e and T are given.

The machine time will represent the true anomaly v, and therefore $\cos v$ and $\sin v$ can easily be generated, by simulating an equation as:

$$\frac{d^2y}{d^2v} + y = 0 . \tag{6}$$

Another advantage of representing the true anomaly v by the machine time, is the fact that the true anomaly is then varying linearly i.e. when the physical value dv/dt is varying fast automatically the physical time passes slowly. As a result of this, the speed of the pen when recording a v_r—t plot is more or less constant also during sharp peaks. In this way, the solution speed, which is limited by the frequency response of the recorder, has been optimised.

In Fig. 4 is given the circuit diagram representing the equations (3), (4), (5) and (6).

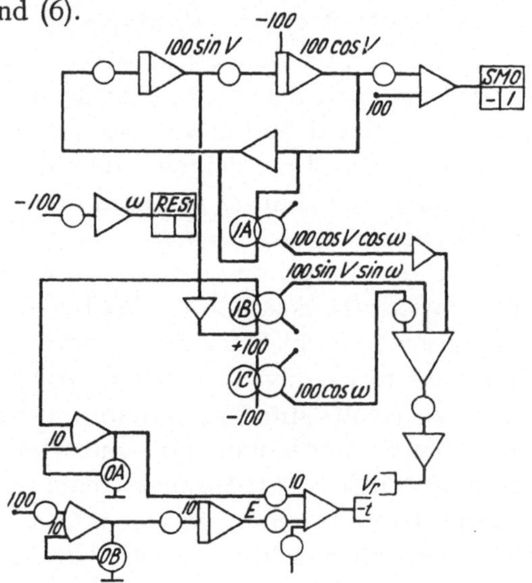

Fig. 4. Circuit diagram representing the equations (3), (4), (5) and (6)

5. Results

A typical result, as it is registrated by a PACE variplotter 1100 E (Fig. 5) is shown in Fig. 1. The abcisse represents time, the ordinate relative radial velocity. As said before, this result is directly plotted on a register card ready for a file.

For the moment, the ephemeris of l'Observatoire Royal de Belgique, consists of about 750 of such cards.

6. Conclusion

At a first glance on the algebraic equations (1), (2) and (3), it seems to be a typical job for a digital machine, to calculate the relative radial velocity of a double star. Looking however in detail and comparing the machine time needed to calculate at descrete points by an iteration process the transcedental equations (1), (2) and (3) on a digital computer, with the 60 secs needed by the analogue computer, the latter has turned out to be in favour. This is even more stressed when taking into account the difference in rental rates for the two types of computers.

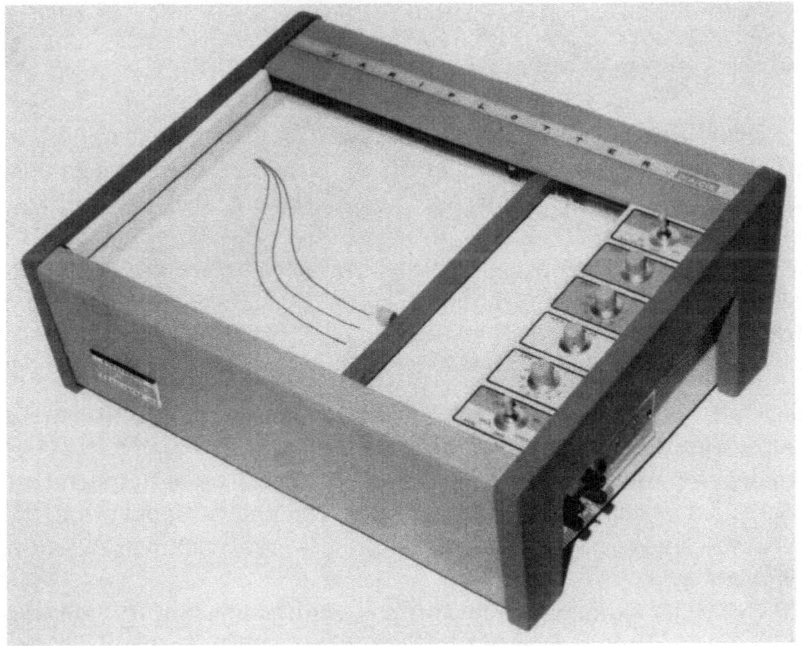

Fig. 5. The PACE variplotter 1100 E

Besides, the representation of the results obtained from a digital computer would be in tables in contrary with the curves plotted by the analogue computer. It should be mentioned that the digital output could be recorded by a dataplotter.

Regarding the accuracy, it should be noted that the calculation of the relative radial velocity at discrete points by a digital computer is more accurate, but on the other hand, this accuracy is lost by interpolating between these points, except in case the calculation is repeated for very many points.

Anyhow, the basic accuracy of the PACE 231-R being 10^{-4}, is more than enough, taking into account that the ephemeris of the relative radial velocity of double stars has to be compared with observations, having an accuracy of the order of 1 km/sec.

30   J. Reynen:

7. Acknowledgement

The author would like to express his gratitude to Dr. J. Dommanget of l'Observatoire Royal de Belgique, who is the originator
of the ephemeris [1], for his permission to present this paper.

References

[1] Dommanget, J.: Communication de l'Observatoire Royal de Belgique
„Catalogue Ephémérides de Vitesses Radiales Relatives Etoiles Doubles
Visuelles". (To be published.)

Discussion

Schwesinger: Mr Reynen erwähnte die Typenbezeichnung des
Analogrechners, mit welchem er die dynamischen Gleichungen aus-
wertete. Dieser Rechner ist mir nicht bekannt, aber ich war sehr
beeindruckt von der hohen relativen Genauigkeit von 10^{-4}, die
Mr. Reynen erwähnte, ich würde gern einige technische Details
erfahren. Handelt es sich wirklich um Genauigkeit oder nur um
Reproduzierbarkeit?

Reynen: The accuracy of a calculation on an Analog Computer
depends on the basic accuracy of the computer as well as on the
calculation technique used. The basic accuracy of a PACE 231-R
Analog Computer is 10^{-4}, which is obtained by a temperature
controlled oven, taking care that the components governing the
accuracy, like resistance and capacitor networks, remain at constant
temperature.

But if, for example, you want to introduce an arbitrary function
into your calculation, using a diode function generator of 20 seg-
ments, it is obvious that the accuracy depends on the shape of
your function. In such eases, I agree, it is better to speak about
reproducibility of 10^{-4}.

Irwin: Does the program refer to visuel binaries to be observed
spectroscopically or to long period spectroscopic binaries?

Strand: This was discussed at IAU Berkeley. The program was
to find the critical times (largest radial velocity differences) to
spectroscopically observe visual binaries, with relatively short
periods $(p < 100 - 200$ yrs.$)$.

Irwin: This would give then, with 2 critical observations, if
the magnitude difference were not too great, both the mass ratio
and the parallax. Unfortunately this is just the type of work
which people with large reflectors do not like to do.

C. Kühne (Berlin-Mariendorf): The Steering-System of a 210-ft Radio-Telescope. (With 9 Figures.)

Zusammenfassung: Es wird an Hand von Lichtbildern das Prinzip der photoelektrischen Nachführung des Radioteleskopes durch ein Leit-Äquatorial (master equatorial) astronomischer Genauigkeit ausführlich erläutert. Die einzelnen Teile und deren besondere Merkmale werden beschrieben.

An Hand eines Blockschaltbildes werden das Zusammenwirken der Baugruppen, sowie das Steuersystem und die verschiedenen Eingriffsmöglichkeiten erläutert.

The following paper reports on some investigations carried out, during the last two years in connection with the building of the large Radio-Telescope for Australia.

The purchaser of this project was the Radio-Physics-Division of the Commonwealth Scientific and Industrial Research Organisation in Sydney. The execution laid in the hands of an English-German team. The elaboration of the plans and the fixing of the essential data of the Radio-Telescope had been entrusted to the Consulting-Engineers Freeman, Fox and Partners in London. The general-contractor, the MAN in Gustavsburg, was responsible for the project in general as also for the erection of the parabolic reflector. The Associated Electrical Industries in Manchester delivered the driving-system for the moving of the Radio-Telescope, inclusive of the installation of the instrument. We, the Askania-Werke in Berlin, assumed the astronomical steering and control-system of the Radio-Telescope.

In general, the reflector of a Radio-Telescope is so heavy that it can only be erected for altazimuthal motion, which means that the reflector is turnig about a horizontal altitude axis, that is mounted on a vertical azimuth-axis. Unfortunately, this altazimuthal mounting does not allow to follow a cosmic object on a simple way, because the cosmic object seems to be turning around an axis, which is parallel to the rotation axis of the earth.

If you nevertheless want to follow celestial bodies with an altazimuthal mounting, you have to calculate position and speed in azimuth and elevation as a function of the star-coordinates, the geographical coordinates of the observation-point and the sidereal time. With regard to the required high accuracy, you need a computer of high resolving power, generally of a digital type. In transmitting these calculated signals into the driving system you have to accept all the unknown erros which are arising in the mounting of the Radio-Telescope.

At first, we chose for this Radio-Telescope another way which renders the coordinate-computer unnecessary as it eliminates all errors of the mounting at once. The method itself goes back to the invention of the English-man Wallis.

Fig. 1 shall explain to you this system.

The Reflector (1) is mounted on the ele-vation axis and balanced by twocounterweights (2). The bearings of this axis are fixed in a steel tower (3), which is rota-table on a circular track (4) on the top of a tower (5). Both axes cut each other in one point (6) in the interior of the steel tower. The steering system now consists to putting a small guiding instrument (7) into this intersection-point (6) and mount it equatori-ally, so that its principal axis directed parallel to the axis of the earth.

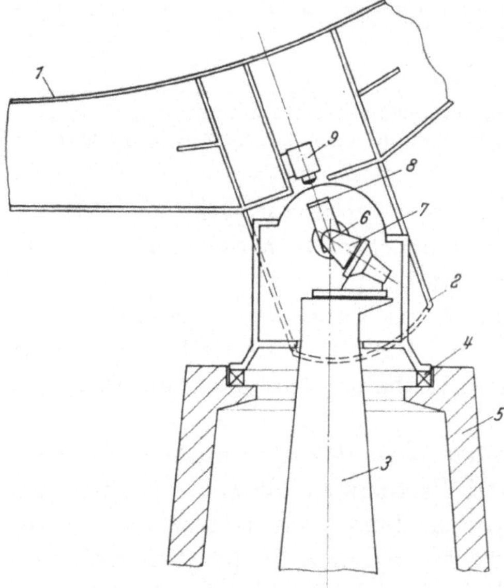

Fig. 1. Principle of the steering system

The small guiding instrument, the so called Master Equatorial, is practically an astronomical instrument of conventional design of highest precision, and carries on the front side of the telescope tube a plane mirror (8). Its normal defines the direction into which the Radio-Telescope shall be turned. Both instruments have a common point of intersection so that the reflector may move around the Master Equatorial.

In the center of the reflector is fastened a so called Error Detec-tor (9) which essentially consists of an autocollimation telescope and 2 photoelectrical devices. They each produce an electrical signal, if the optical axis of this Error Detector does not stand exactly vertical on the mirror of the Master-Equatorial, and they produce each a signal for the misalignement in azimuth an elevation direc-tion. With the help of calibration measurements the electrical axis of the reflector is aligned parallel to the optical axis of the Error Detector. After a suituable processing both error signals are used for the control system of the drive motors so that the electrical axis of the reflector is tilted always exactly parallel to the direction given by the mirror of the Master-Equatorial. By this method all errors in the mounting are eliminated at once.

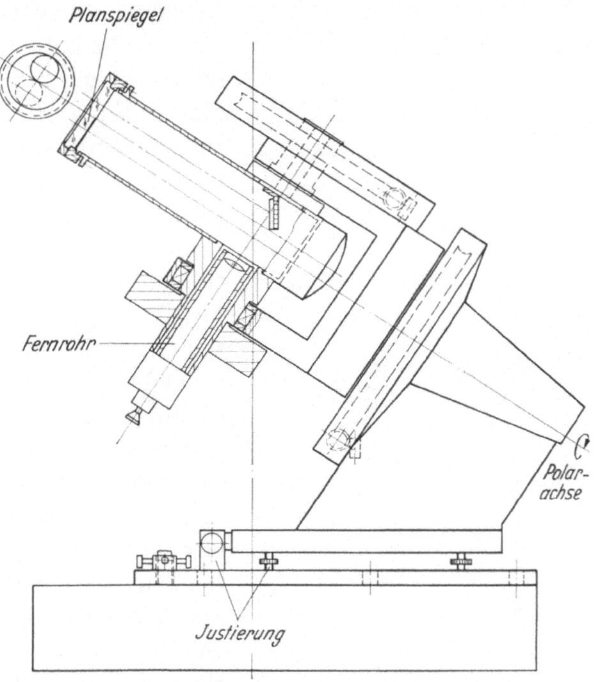

Fig. 2. The Master-Equatorial

The greatest part of the deformation errors, namely all which are rotation-symmetrical and proportional to the sine of the zenith angle will be removed by a simple but rather effective elastic compensation in the suspension of the Error Detector. Practically there only remain the deflections of the ellipticity of the paraboloid, but it is possible to keep them small by an especial design of the reflector surface.

Fig. 2 shows the principle of the design of the Master-Equatorial. You see an astronomical instrument of a so called fork type. Its polar axis is inclined 33 degrees, corresponding to the geographical latitude of the observation station. Both axes, the hour-angle and the declination axis are driven each by a wormwheel of highest precision. Colinear to the declination-axis a telescope has been placed which serves for adjusting purposes, and which allows to observe the sky with the help of a 45°-mirror and through the unsilvered part of the main-mirror.

Fig. 3 represents the principle of the Error Detector. With the help of the objective an illuminated screen will be imaged through halfsilvered plates to infinity. The light than falls on the plane mirror of the Master-Equatorial and is reflected into itself, if this mirror is exactly vertical to the optical axis. The reflected beam attains over the half-silvered plates 2 dividing edges of the optical system, which has been arranged so, that these 2 dividing edges are vertical to each other in the field of view of the reflected beam and represent a coordinate system, namely the azimuth and the zenith angle. The two light beams will be split into couples by the dividing edges and reach via a rotating shutter a photo-multiplier each. In the following electronic arrangement they will be compared photometrically. The result of the photometry is a signal, which is proportional to the difference of both split beams, respectively, referring to the dividing edges of the field of view. The signal disappears, if the reflected screen-image is in the origin of the coordinate system.

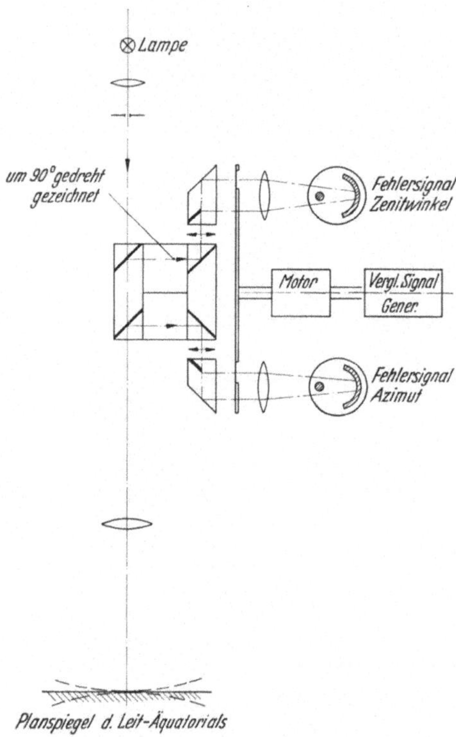

Fig. 3. Principle of the Error Detector

In Fig. 4 you recognize the arrangement of the Master-Equatorial and the Error Detector together with a panel. The Error-Detector looks through the hole in the steel-ceiling.

In designing the steering system it was to consider that the steering should ensue from a room laying inside of the tower which does not allow a direct observation of the instrument. For the observer to be able to see the position of the telescope, we built a small coordinate-model, the so called Slave-Model, where it is possible to recognize the position of the reflector and the Master-Equatorial with an accuracy of about $\frac{1}{2}$ degree. Fig. 5 shows the principle of this model. We put in the centre a small equatorial imitation. Its pointer possesses the direction defined by the Master-Equatorial. The vertical axis is copied in the foot of the model and the horizontal altitude-axle at the left arm. Its pointer represents the direction of the reflector. Additionally, a compass card is fixed which shows the wind direction. The Slave-Model is covered by a sphere of plexiglass in which the altazimuthal and equatorial coordinate-nets are engraved. Besides the general orientation, it is especially the task of this model to bring both systems so good in coincidence at the beginning of a measurement, that the Error-Detector will approximately be brought in opposition to the mirror

Fig. 4. Arrangement of the Master-Equatorial and the Error Detector together with a panel

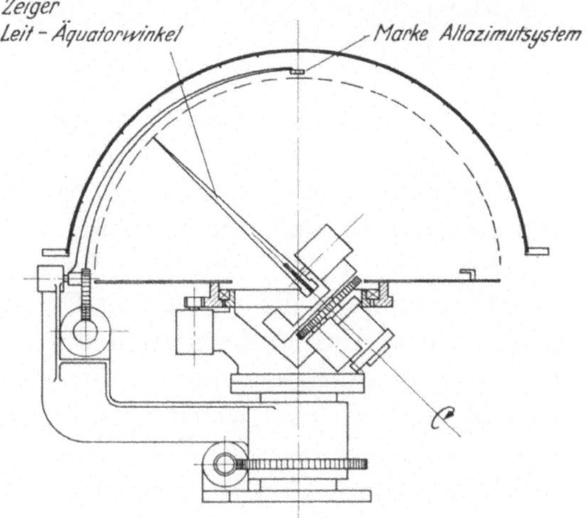

Fig. 5. Principle of the coordinate-model

of the Master-Equatorial, because its field of view is about plus and minus one degree only.

Furthermore the steering comprises all gearing and intervention possibilities into the system, the necessary controls, the measuring devices for all coordinates and two quarz-clocks with time-setting-system. The principle is represented in the block-scheme of Fig. 6.

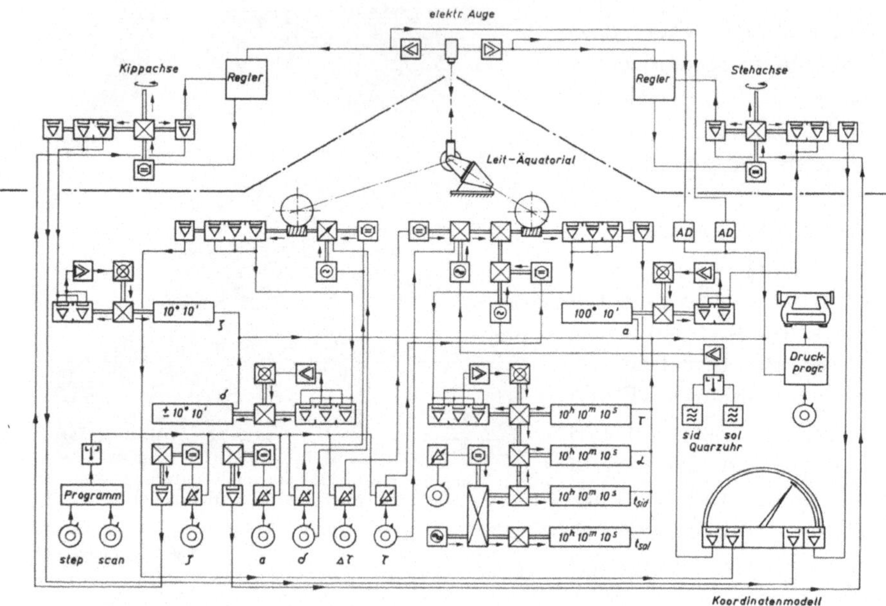

Fig. 6. Block-scheme of the steering and control system

Above the thick line you see all parts which are connected to the reflector, on the left the elevation axle, on the right hand the azimuth-axle, in the centre the Error Detector. Under the line you find symbolically the Master-Equatorial with its two worm-wheels and worms for declination and hour-angle. The hour-axis is driven by four different motors. First a synchronous motor for the normal celestial rate which gets its frequency changed by the sidereal or solar quarz-clocks, second, a rate-controlled dc-motor which superimposes a very small rate $(\Delta\tau)$ of about 4% to the celestial rate. This rate gives the difference rate to follow a planet. Third, a rate-controlled dc-motor for scanning (τ) the hour-axis. The scanning rate is continuously selectable by a dial between 0.02 and 2.5 deg/min. At last, an ac-motor for rough slewing of the hour-axis up to 60 deg/min. The declination axis has also the scanning (δ) and slewing motors, but no celestial motors.

To measure the positions in hour-angle and declination with high accuracy each worm-axis is connected to a three-step magslip-arrangement. In the control desk we have a similar receiver arrangement with a follow-up servo-system. This device comprises an analog-digital-converter which translates the transmitted values into numbers. The smallest digital unit is 0.1 min of arc in declination (δ) and 1 sec of time respectively 15 sec of arc in hour-angle (τ). The hour-angle arrangement is extended by similar digital units to show the right ascension (α) and the sidereal (t_{sid}) and solar (t_{sol}) time. The computation is accomplished by a similar gear-mechanism driven by a synchronous motor in connection with the solar quarz-clock and a time correction-system for periodical time-checking.

To measure the angles in azimuth and elevation we use also magslips and follow-up-system but with 2 steps only, because the required accuracy is lower. Furthermore, the altazimuth-axes are direct movable by so called velodynes with rates up to 24 deg/min in azimuth (α) and 15 deg/min in elevation (ζ).

A program-unit makes possible to scan automatically the sky direct in altazimuth coordinates or in equatorial coordinates. In the last case and during the celestial motion the Error-Detector takes over the guiding of the altazimuth-axis. The residual errors in azimuth and elevation represented by the amplitudes of the errors signals caused by wind-pressure and other unequalities are also transmitted in digital values by analog-digital converters.

An automatical print-out-system feeds all the essential numbers into an IBM-typewriter which writes the full sequence in open, uncoded numbers in selectable intervals down to 10 seconds. The Slave-Model gets its signals from magslips-system of similar types like the other.

The steering arrangement itself has been built into a Control-Desk and two Remote-Panels.

Fig. 7 shows the Control-Desk. You see on the left hand the Slave-Model, the intercommunication-system and some control lamps with warning system. In the central part you see the digital position-indicators for azimuth, zenith-angle, hour-angle, declination, right ascension, sidereal and solar time, the residual error-indicators, rate indicators, time-setting-system and the knobs for

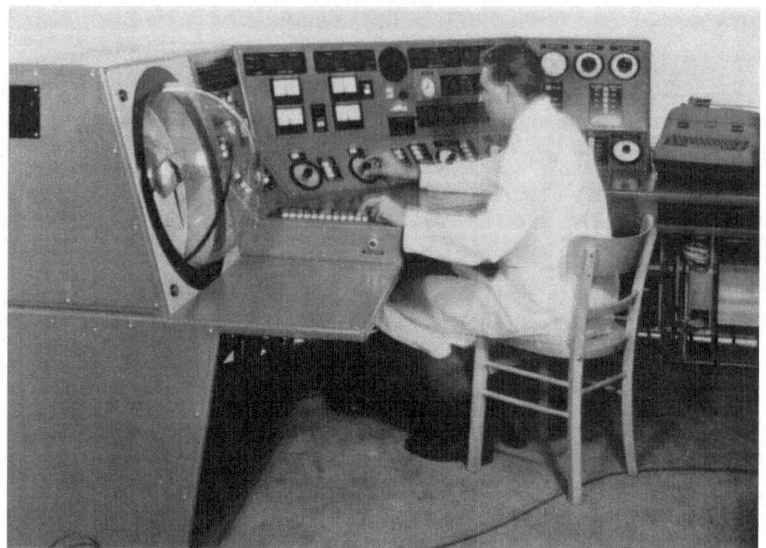

Fig. 7. The Control-Desk

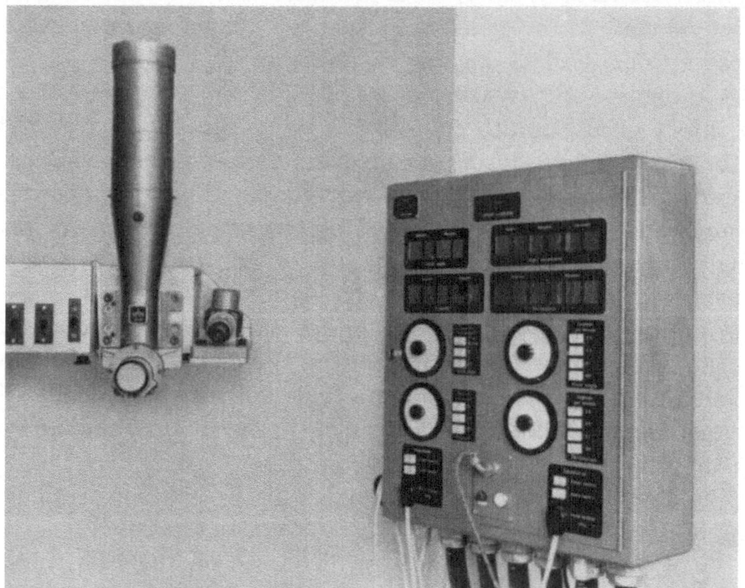

Fig. 8. One of the Remote-Panels

the different motions of the telescope. On the right hand, you see
the knobs to select the scan-program and the print-out-system
with a typewriter.

Fig. 8 shows one of the Remote-Panels which is mounted in the so called vertex-room, direct under the surface of the reflector. The Remote-Panel allows some steering possibilities, special motions to follow a noncelestial body by optical observing with a good optical telescope on the left hand.

Fig. 9. The 210-ft Radio-Telescope of the Radio-Physics Division of the Commonwealth Scientific and Industrial Research Organisation in Sydney

Fig. 9 shows the whole Radio-Telescope, which is told to be the most modern of its kind. Because of its outstanding guiding precision and its extreme resolving power we may surely expert some interesting investigation results.

Discussion

Behr: Are there any provisions made to rotate the antenna according to the relative rotation of the horizontal system with respect to the equatorial system in order to avoid errors which may arise at polarization effects?

Kühne: This is unknown to me because whole the electronic equipment is an own development of the radio-astronomers of the C.S.I.R.O. It may be that such a rotation is possible.

H. Plesse (Oberkochen): Photoelektrische Fernrohrnachführung
für Sonnenbeobachtung. (Mit 5 Textabbildungen.)

Zusammenfassung: Das photoelektrische Sonnenleitfernrohr ist eine
selbständige Baueinheit, die an beliebige Einrichtungen zur Sonnenbeob-
achtung angebaut werden kann. Sie ist ursprünglich für den 150 mm-
Coudé-Refraktor von Carl Zeiss, Oberkochen, entwickelt worden.
Das Gerät arbeitet im Prinzip nach der Einzellen-Wechsellicht-Methode.
Dabei wird für jede Koordinate jeweils aus zwei einander gegenüberliegenden
Teilen am Rand des Sonnenbildes Licht mittels Kreisring-Sektoren ausge-
blendet, mit Hilfe einer rotierenden Halbkreisscheibe abwechselnd abgedeckt
und freigegeben und dann einer Photozelle zugeführt. Bei unsymmetrischer
Lage des Sonnenbildrandes zu den Sektoren gelangt in jeder Halbperiode je
nach Verschiebungsrichtung unterschiedlich viel Licht auf die Photozelle.
Es entsteht ein Photowechselstrom, der nach Verstärkung über einen Servo-
motor das Leitfernrohr zusammen mit dem Beobachtungsinstrument so
lange bewegt, bis die symmetrische Lage, die gleichzeitig die Sollage des
Sonnenbildes in bezug auf die Sektorblende darstellt, erreicht ist. In dieser
Lage verschwindet die den Motor treibende Wechselspannung. Der Rektaszension wird eine gleichmäßige Bewegung mittels Synchronmotor überlagert.
Der Fangbereich des Leitfernrohres beträgt in beiden Koordinaten 30 Bogenminuten. Die Genauigkeit der Nachführung ist besser als 1 Bogensekunde.

Für einen 150 mm-Coudé-Refraktor wurde eine photoelektrische
Fernrohrnachführung entwickelt, die nach dem Prinzip der Zwei-
strahl-Einzellen-Wechsellichtmethode arbeitet. Diese Nachführeinrichtung, auch Sonnenleitfernrohr genannt, ist eine selbständige
Einheit, die an verschiedene Sonnenbeobachtungsgeräte angebaut
werden kann. Voraussetzung ist lediglich die Anpassung der End-
stufen der Verstärker und der Motoren an den jeweiligen Leistungs-
bedarf für die Bewegung.

Das Schema der Meßeinrichtung, mit deren Hilfe die Lage des
Sonnenbildes in bezug auf die Fernrohrachse festgestellt wird, ist in
Abb. 1 dargestellt. Das 50 mm-Objektiv L_1 bildet gemeinsam mit
dem Projektiv L_2 die Sonne in die Ebene einer vierteiligen fest-
stehenden Sektorenblende Bl entsprechend dem darunter ange-
gebenen Schnitt ab. Der Mittelpunkt der Blende liegt in Fernrohrachse.

Der innere Durchmesser der Sektorenblende ist um 4 % kleiner
als der Durchmesser des Sonnenbildes, die jeweilige Bogenlänge der
vier Kreisringsektoren beträgt $^1/_{12}$ des Kreisumfanges.

Das Licht von jeweils zwei einander gegenüberliegenden Sektoren, die paarweise der Rektaszension α oder der Deklination δ
zugeordnet sind, wird mit Hilfe von Umlenkprismen je einer

Photozelle Ph zugeführt. Dabei bilden Projektiv L_2 und Feldlinse L_3 die Objektivöffnung auf die Photozellen ab, so daß die Empfängerflächen gleichmäßig ausgeleuchtet sind.

In unmittelbarer Nähe der Sektorenblende rotiert konzentrisch zu ihrer Achse eine Halbkreisblende S, die von einem Synchronmotor mit gleichbleibender Phasenlage gegenüber der Netzspannung angetrieben wird. Diese Halbkreisblende deckt den Strahlengang der einander gegenüberliegenden Kreisringsektoren

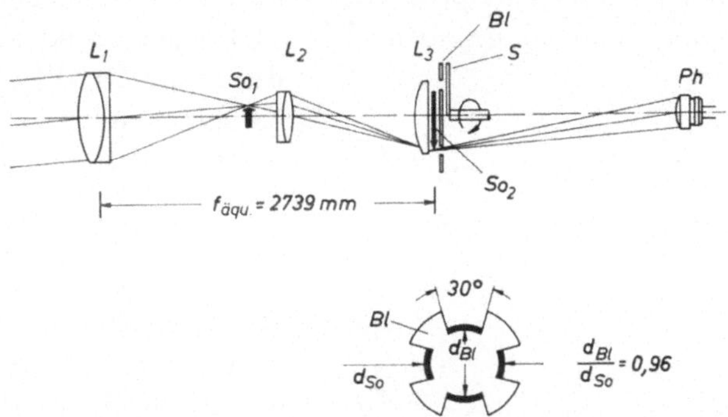

Abb. 1. Optisches Schema des photoelektrischen Sonnen-Leitfernrohres

im Wechsel mit Netzfrequenz 50 Hz ab, wobei die Abdeckung der beiden räumlich um 90° versetzten Sektorenpaare mit einem Phasenunterschied von ebenfalls 90° erfolgt.

Die Bezugslage des Sonnenbildes für die Fernrohrnachsteuerung ist die Symmetrielage zur Sektorenblende. In diesem Fall fällt nämlich auch bei Rotation der Halbkreisblende nur Gleichlicht auf die Photozellen und zwar im Wechsel von einer der beiden gegenüberliegenden Kreisringsektoren. Die dabei erzeugte Gleichspannung wird durch den angeschlossenen Wechselspannungsverstärker nicht verstärkt. Bei Verschiebung des Sonnenbildes in Richtung der Verbindungslinie eines Sektorenpaares, ändert sich der Lichtstrom in jeder Halbperiode, und zwar ist das Vorzeichen der Verschiebung maßgebend für die Phasenlage der nunmehr entstehenden Wechselspannung. Die Vorgänge in beiden Koordinaten entsprechen einander, für eine Koordinate sind die Licht- und Spannungsverhältnisse in Abb. 2 dargestellt.

Das obere Bild zeigt den Verlauf der Lichtströme als Funktion der Zeit, man erkennt den Übergang der Lichtstromanteile aus den beiden zugeordneten Kreisringblenden beim Durchgang der Kante der Halbkreisscheibe durch die Sektoren. Die halbe Differenz der Lichtstromamplituden ist ein Maß für die entstehende Photowechselspannung.

Unten ist die Einhüllende der Wechselspannung bei Ablage des Sonnenbildes unter Berücksichtigung ihrer Phasenlage dargestellt. Beim Durchgang der Spannungsamplitude durch Null in der Bezugslage des Sonnenbildes wechselt ihre Phase von Null nach 180°. Das würde bei Gleichspannung dem Wechsel der Polarität von Plus nach Minus entsprechen. Dadurch ist eine eindeutige Zuordnung der Richtung gegeben. Die bei Ablage $\varDelta\alpha$ oder $\varDelta\delta$ von der Symmetrielage an den entsprechenden Photozellen entstehende Spannung treibt nach Verstärkung für jede Koordinate einen Nachlaufmotor in solcher Richtung an, daß durch die resultierende Bewegung des Fernrohres die Spannungen bis auf einen geringen Rest zu Null werden. Maßgebend für die Genauigkeit der Nach

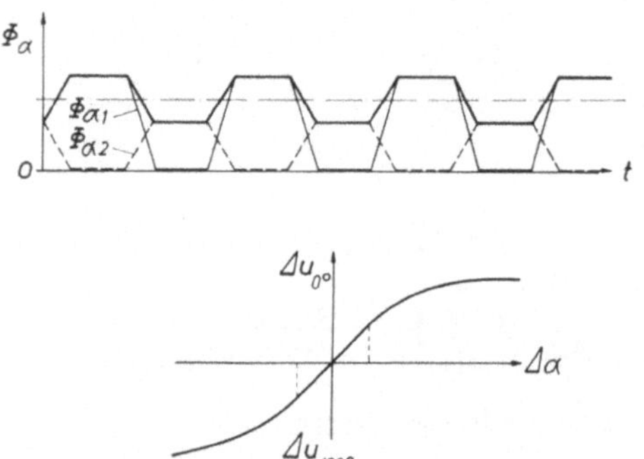

Abb. 2. Lichtstrom und resultierende Wechselspannung bei Ablage

steuerung ist der Anstieg der Spannung im Nulldurchgang. Je steiler der Anstieg, um so größer die Genauigkeit, vorausgesetzt, daß nicht infolge zu großer Steilheit Regelpendelungen, die die Grenze der Verstärkungsmöglichkeit bilden, auftreten. Um eine einmal eingestellte optimale Regelgüte, deren Maß das Produkt aus Regelgeschwindigkeit und Einlaufgenauigkeit ist, aufrechtzuerhalten, ist es notwendig, den Spannungsanstieg am Motor in Abhängigkeit von der Ablage auch unter veränderten äußeren Bedingungen, z. B. bei erhöhter Absorption der Sonnenstrahlung durch die Atmosphäre, beizubehalten.

Bei der hier angewendeten Nachlaufregelung handelt es sich um ein Nullverfahren, das in besonders hohem Maße unempfindlich gegen äußere Einflüsse ist. Änderungen des Verstärkungsgrades etwa durch Netzspannungs- oder Netzfrequenzänderungen, ferner Änderungen der Photozellencharakteristik sowie Nichtlinearitäten des Verstärkers gehen in weiten Grenzen nicht in die Genauigkeit der Einstellung des Sonnenbildes auf die Bezugslage ein.

In Abb. 3 ist die Gesamtanordnung entsprechend der praktischen Verwirklichung perspektivisch im Schema dargestellt. Man erkennt daraus die räumliche Anordnung. Das Licht wird nach mehrmaliger Reflexion an den Umlenkprismen in die Einfallsrichtung zurückreflektiert und fällt auf die auf einem Ring in der Nähe des Objektivs befindlichen Photozellen. Dadurch wird die Gesamtbaulänge erheblich verkürzt, sie beträgt nur 750 mm.

Die Abschätzung der Energieverhältnisse ergibt bei einer Öffnung des Fernrohres von 50 mm freiem Durchmesser einen Ge-

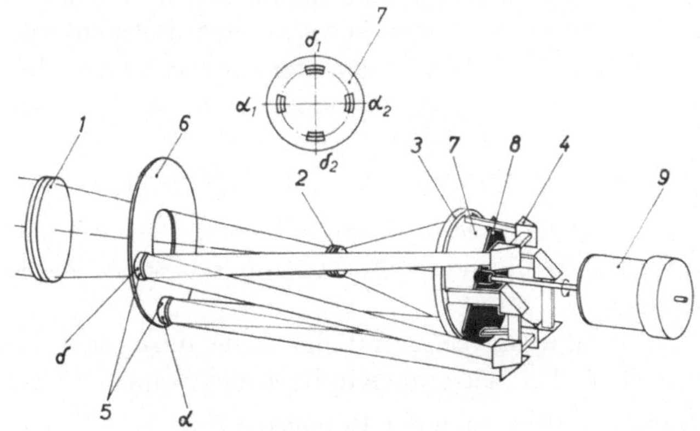

Abb. 3. Schematische Darstellung der Gesamtanordnung des photoelektrischen Sonnen-Leitfernrohres

samtlichtstrom von rund 70 lm. Hiervon geht bei Symmetrielage durch einen Kreisringsektor unter der Voraussetzung, daß das Verhältnis von Durchmesser der Kreisblende zu Sonnenbilddurchmesser 0,96 beträgt und daß $^1/_{12}$ des Gesamtumfanges ausgenutzt wird, ein Anteil von $6,6 \cdot 10^{-3}$ hindurch. Damit stehen zur Messung 0,46 lm zur Verfügung, die der Photozelle zugeführt werden. Die Beleuchtungsstärke bei voll ausgeleuchteter Photozelle mit 10 mm Durchmesser wird damit $E_z = 5,8 \cdot 10^3$ lx.

Einer Verlagerung des Sonnenbildes um $1''$ in der Blendenebene entspricht eine Linearverschiebung von 13 μm. Das ergibt eine relative Flächenänderung $\Delta F/F$ in der Sektorenblende von $2,6 \cdot 10^{-2}$, und somit eine Änderung der Beleuchtungsstärke um rund 150 lx. Das sind Energiewerte, die die Verwendung eines Photoelementes als Empfänger als völlig ausreichend erscheinen lassen, ohne daß bereits ein Rauschen des Empfängers oder des Verstärkereingangs merklich zu erwarten ist.

Es war bereits darauf hingewiesen worden, daß zur Aufrechterhaltung einer optimalen Regelgüte die Spannungsänderung am
Motor im Nulldurchgang unabhängig von äußeren Bedingungen,
z. B. also auch von der Beleuchtungsstärke, konstant gehalten
werden muß. Man macht das im allgemeinen mit verhältnismäßig
viel Aufwand dadurch, daß man mit Hilfe der einfallenden Energie
die Verstärkung automatisch regelt, ähnlich wie z.B. beim Schwundausgleich bei Rundfunkempfängern. In unserem Fall nun kommt
der genannten Forderung die Charakteristik der Photoelemente
entgegen, d. h. die Abhängigkeit der Photospannung vom auffallenden Lichtstrom oder der herrschenden Beleuchtungsstärke.

Bekanntlich ist bei Photoelementen der Kurzschlußstrom I_K
weitgehend proportional zur Beleuchtungsstärke E, während die
Photospannung U bei geeigneter Belastung, und zwar annähernd
im Leerlauf, in einem weiten Bereich proportional zum Logarithmus
der Beleuchtungsstärke ist. Es gilt also $U = U_0 \cdot \log \dfrac{E}{E_0}$. Durch
den Spannungswert U_0 und die Beleuchtungsstärke E_0 ist die Kennlinie festgelegt.

Die Beleuchtungsstärke E auf der Zelle möge sich von einer
Ausgangsgröße E_1 bei Sektorausschnitt α_0 des Sonnenbildes mit der
Verschiebung α ändern nach der Beziehung $E = E_1\left(1 + \dfrac{\alpha}{\alpha_0}\right)$. Dann
wird

$$U = U_0 \log \frac{E_1\left(1 + \dfrac{\alpha}{\alpha_0}\right)}{E_0},$$

und die Ableitung nach α, die ein Maß für die Empfindlichkeit ist,

$$\frac{dU}{d\alpha} = \frac{U_0}{2,3} \frac{1}{\alpha + \alpha_0}.$$

Man erkennt, daß diese abhängig ist von der ursprünglichen
Größe des Sektorbereiches α_0, aber unabhängig von der Beleuchtungsstärke E. Messungen zeigen, daß die Nachlauf- und Einlaufgenauigkeit von Schwankungen der Beleuchtungsstärke in einem
Verhältnis von $1:20$ nicht beeinflußt wird.

Abb. 4 gibt das Schema der gesamten elektronischen Nachführeinrichtung wieder. Die Photoelemente Ph_1 und Ph_2 links oben
im Bild liefern die Steuerspannungen, der Synchronmotor M_1,
dessen Achse phasenstarr mit der Netzspannung läuft, trägt die
Halbkreisblende zur Unterbrechung der Strahlung. Die Steuerspannungen werden zunächst in den Vorverstärkern V_1, an-

schließend in den Leistungsverstärkern $V_{2/3}$ für jeden Kanal so weit verstärkt, daß sie für den Antrieb der Nachlaufmotoren M_3 und M_4 ausreichen. Statt der Photozellen und Vorverstärker können mit Schalter S_1 wahlweise auch die Fernsteller $F\,S_{1/2}$ an die Leistungsverstärker geschaltet werden. Durch die Betätigung von Potentiometern ist so auch eine Steuerung in beiden Koordinaten von Hand möglich. Bei den Motoren $M_{3/4}$ für die Bewegung in Rektaszension α und Deklination δ handelt es sich um Zweiphasen-Induktions-

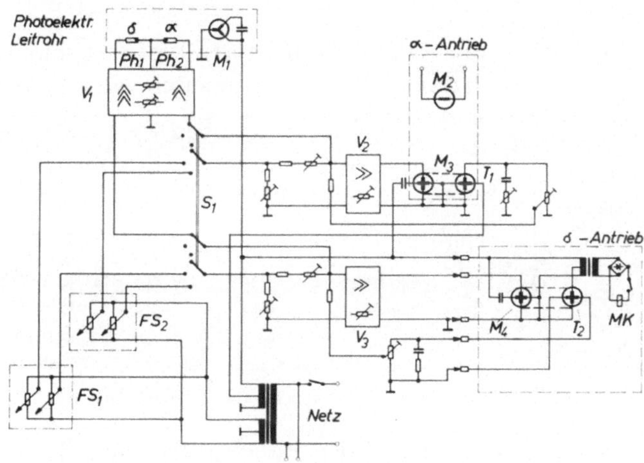

Abb. 4. Schaltschema der elektronischen Nachführeinrichtung

Motoren. Ein solcher Motor erhält eine feste Hilfsspannung vom Netz und eine um 90° dagegen versetzte Steuerspannung vom Verstärker. Durch die Überlagerung der beiden magnetischen Felder dieser Spannungen kommt ein Drehfeld zustande, das den Kurzschlußläufer, in dem Ströme induziert werden, mitnimmt. Je nach Phasenlage der beiden Spannungen zueinander dreht sich der Motor in der einen oder anderen Richtung auf den Abgleichpunkt zu. Mit den Motoren sind sog. Tachogeneratoren direkt verbunden. Der Name rührt daher, daß sie eine zur Geschwindigkeit der Motorachse proportionale Spannung erzeugen. Diese Spannung wird mit einer der Regelspannung entgegengesetzten Phase dem Verstärker zugeführt. Es handelt sich also um eine Maßnahme, die man in der Verstärkertechnik als Gegenkopplung bezeichnet. Sie gibt die Möglichkeit, die Übertragungseigenschaften eines Verstärkers bezüglich Linearität und Stabilität erheblich zu verbessern. Hier in einer Regelschaltung verbessern sie als differenzierende Rück-

führung die Nachlaufeigenschaften in bezug auf Einlaufgeschwin-
digkeit und Dämpfung. So werden z. B. die Auswirkungen von
mechanischer Lose und unterschiedlicher Haft- und Bewegungs-
reibung, die bekanntlich bei Regelanlagen die Schwingneigung
erhöhen, ganz wesentlich verringert und somit die Regelgüte verbessert. Der Synchronmotor M_2 treibt die Stundenachse an. Mit
Magnetkupplung M K kann die Verbindung zwischen Getriebe und
Säule gelöst werden, so daß eine Verstellung des Fernrohrs von
Hand möglich ist.

Die Verstärker, die voll mit Transistoren bestückt sind, sind zur
leichten Austauschbarkeit als Steckeinheiten ausgebildet. Sie

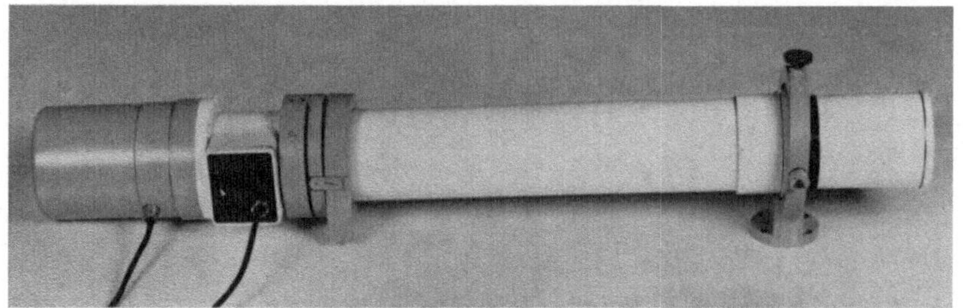

Abb. 5. Das Sonnen-Leitfernrohr im abmontierten Zustand

können wegen der Kleinheit der Bauelemente und der geringen
Leistungsaufnahme klein gehalten werden, Gesamtvolumen beträgt
etwa 2500 cm³. Das Verstärkergerät ist beim Coudé-Refraktor in
der Säule eingebaut und über Kabelleitungen von einigen Metern
Länge mit der Meßeinrichtung verbunden.

Abb. 5 zeigt das Sonnenleitfernrohr im abmontierten Zustand.
Die Montierung wird am Kardangelenk und an der Justiereinrichtung vorgenommen. Unter der Haube am unteren Ende befindet
sich der Synchronmotor zum Antrieb der Halbkreisblende. Am
Schaltbrett befindet sich noch ein Schalter zur Phasenumkehr der
Photozellenspannung für Bewegung in Rektaszension. Diese Um-
schaltung wird notwendig bei Umkehrung der Drehrichtung infolge
veränderter Fernrohrstellung. Durch die Justiereinrichtung ist es
möglich, jeden beliebigen Punkt der Sonne in die optische Achse des
Fernrohres und damit des Gesichtsfeldes zu legen. Die Photozellen
sind nach Zurückschieben eines Ringes in der Nähe der Justierung
zugänglich.

Die praktische Erprobung des Sonnenleitfernrohres an einem
Coudé-Refraktor 150 mm ergab eine statische Einlaufgenauigkeit

von mindestens einer Bogensekunde. Auf diese Genauigkeit konnte auf zweierlei Weise geschlossen werden, nämlich einmal durch Beobachtung der Winkeleinstellung der Motorachse nach zwangsweiser Auslenkung des Gerätes und ferner durch Kontrolle der Spannung am Verstärkerausgang bei normalem Nachlauf. Beide Größen lagen stets merklich unterhalb der Werte, die einer Bogensekunde Abweichung entsprechen würden. Selbstverständlich besagt das nur, daß mit dieser Genauigkeit auf Strahlungsgleichgewicht zwischen den beiden jeweils gegenüberliegenden Sektorausschnitten eingestellt wird. Der Fangbereich beträgt bei klarem Himmel 30 Bogenminuten, er entspricht dann also praktisch dem Sonnendurchmesser. Bei dunstigem Himmel kann er wesentlich größer sein.

Im Bereich von rund 3 Bogenminuten bis zu einigen Bogensekunden Ablage des Sonnenbildes vom Bezugspunkt läuft das Fernrohr mit einer Maximalgeschwindigkeit von 100 Bogensekunden pro Zeitsekunde, innerhalb von etwa 5 Bogensekunden dann proportional zur Ablage. Die Gefahr eines Herauslaufens aus dem Fangbereich scheint nach den gemachten Erfahrungen auch beim Vorhandensein von Wolkenfeldern gering zu sein.

Diskussion

Høg: Wie groß sind Zeitkonstante und Korrekturgeschwindigkeit?

Plesse: Die maximale Korrekturgeschwindigkeit beträgt $100''/$ sec. Die Zeitkonstante ist praktisch allein durch den Nachlaufmotor gegeben, da die Belastung infolge der hohen Übersetzung vernachlässigbar klein ist. Die Beschleunigungszeit vom Stillstand bis 90 % der Leerlaufdrehzahl beträgt 0,03 sec. Das gilt in dem Spannungsbereich, in dem die Drehzahl proportional zur angelegten Spannung ist.

E. Høg (Hamburg-Bergedorf): The Photoelectric Micrometer for the Bergedorf Meridian Circle. (With 4 Figures.)

Summary: A photoelectric micrometer for the meridian circle will be ready for testing in August 1962. The micrometer mainly consists of a fixed focal grid and a pulse counting photometer with digital output on punched tape. The indication of the counter is punched every 100 ms of time on the tape from which a computer can later deduce the transit time and also the declinationcomponent. Using the known power spectrum of seeing it has been shown that the accuracy decreases very little for stars of high declination although at 84° of declination only 3 slits are crossed during an observing time of 20 seconds.

We will be able to measure stars down to about 10th mag., Sun, Moon, and planets, and the brightest stars also at daytime. The micrometer permits the measurement of collimation, nadir, and of mire.

We will try to read the declination circle (with 3' divisions) by making a microphotometry of 6' of the circle projected onto reference lines. This registration will be done on the punched tape after the observation of each star.

A photoelectric micrometer is under construction for the Bergedorf observatory and will be ready for testing this year. A glass plate with an evaporated grating of about one hundred slits is fixed in the focal plane of the transit instrument. The star being observed crosses these slits one after the other, and the light goes to a photomultiplier behind the grating. The multiplier is the first part of a pulse-counting photometer with digital output every 0.1 sec as governed by an accurate quartz clock and timer. The digital output on the paper tape can later be processed by a digital computer and gives the transit times for each slit. See also Høg [1].

It will be possible to measure stars down to about 10th magnitude, sun, moon and planets, and the brightest stars also at daytime. The micrometer also permits the photoelectric measurement of collimation, nadir and mire, and is therefore well suited for absolute meridian observations.

The declination circle with 3' divisions will be read photoelectrically too by making a microphotometry of 6' of the circle projected onto reference lines. This registration will be done on the punched tape after the observation of each star.

Description of the Micrometer

The mechanical parts of the micrometer are shown schematically at the top of Fig. 1, and are the same as in an ordinary photoelectric photometer except that in the focal plane we have a complicated grating (or grid) instead of a simple circular diaphragm.

The grating to be used at the test on the transit instrument is shown in Fig. 2. It is different from the one to be used at the transit circle, since at the latter also collimation, nadir, mire and

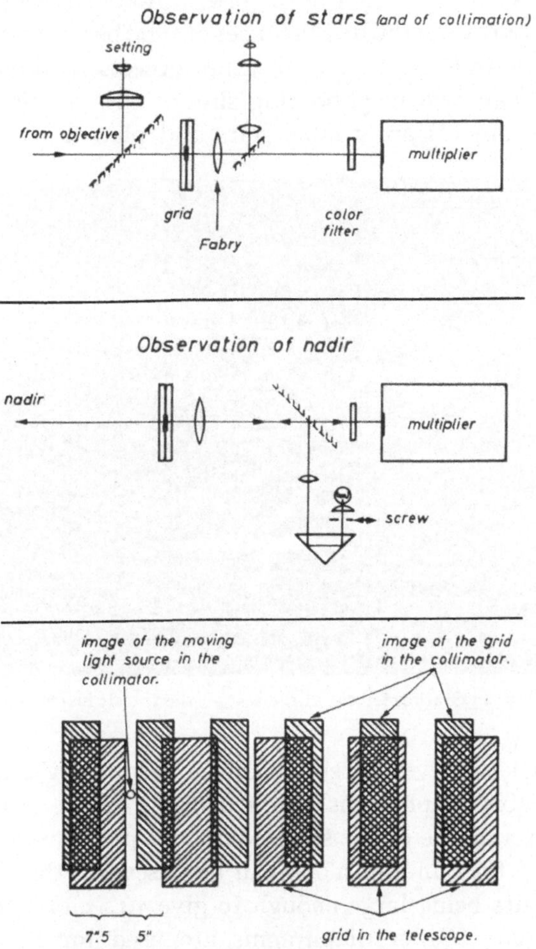

Fig. 1. Scheme of the mechanical parts of the micrometer

declination are measured photoelectrically, but at the former the collimation is measured by reversing the instrument at each star; The inclination is measured by a level. The grating shown consists of five horizontal parts to be used separately. The middle part consists of 90 vertical slite of 5″ width and of one inclined slit at both ends. These latter slits serve for determining the ordinate of the star during transit, and especially, when the ordinate is measured at both ends, for determining the orientation of the whole grating. The second part from top has distances of 100″ between the slits which are 50″ high and are used to measure objects so broad that they appear through two neighbouring slits of the middle part at one time. Such objects are double stars and planets showing a disk.

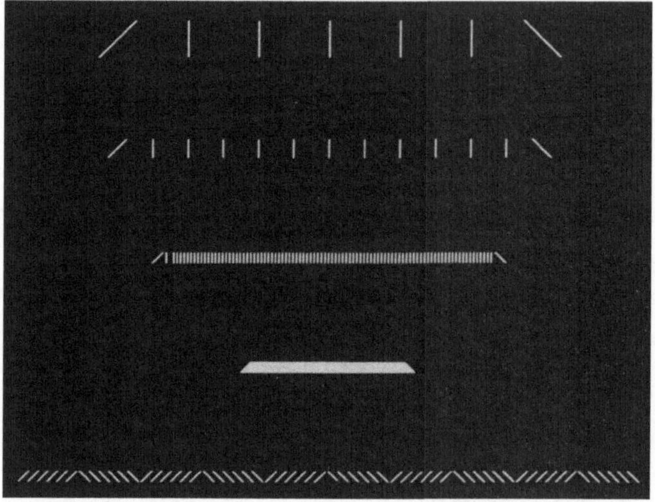

Fig. 2. Grating for use at the test on the transit instrument

The top part has even larger distances between the slits so that interchangeable diaphragms 3 mm in front of the grating permit to block out any one of the slits. With this part daytime stars and the edges of the sun and moon can be observed, the distances between the slits being large enough to give 10 sec of time to change the diaphragm. Other diaphragms are used for the four lower parts of the grating. When very bright objects are observed neutral filters are inserted in front of the multiplier, and for the sun a second filter in front of the objective, so that the sensitivity of the pulse counter is matched. The second part from the bottom of the grating is transparent and permits the measurement of the changing intensity of a star, thus the power spectrum of scintillation. The grid at the bottom is similar to the one to be used in the transit circle, since it can measure also the ordinate of the path of the star. Here at the transit instrument it only serves for studying the later application in the transit circle.

Let us start the detailed description of the electronic equipment by calling it a photoelectric photometer for astrometry, and by stating that it is a pulse-counting photometer with digital output and with a very accurate timer, see Fig. 3. For the pulse-counting parts it need only be said that the linear amplifier apply delay line clipping and that the counter is a 1 Mc/s. All electronics except the amplifier and the high voltage is transistorized. A switch to be set according to the stellar magnitude before each transit selects the three consecutive digits of the counter which will be punched on the tape every 100 ms. The start of the measurement is decided by comparing the right ascension of the star on the observing list with the clock showing sidereal time. Then the start button is pressed. This does not mean the immedeate start of the punching, but it means that at the next following 5 sec pulse the punching starts. The punching stops automatically when the preset is reached, for instance after 60 sec. Now the observ   has to punch manually the star number and the approximate time, ± 2 sec, for the start. The observation of the next star can begin.

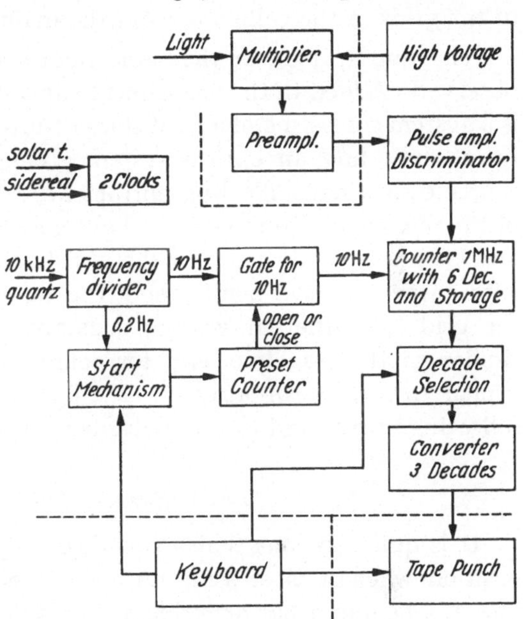

Fig. 3. Photoelectric Photometer for Astrometry. Simplified diagram

Observation of Collimation and Nadir

At the transit circle collimation and nadir are observed photoelectrically as shown in Fig. 1. The lower part of this figure shows the image in the focal plane of the telescope when collimation is being measured. Behind the grid in the collimator a light source can be driven across with constant velocity. Obviously, the multi-plier only receives light when neither the grid in the collimator nor that in the main telescope cut off the light, and further that the total amount of light passing each such slit is proportional to its width. Therefore the computer will be able to evaluate the punched photometric data in order to obtain the relative position of the grids in the collimator and the main telescope.

The middle part of the same figure shows how nadir can be observed. Added to the mechanical parts shown at the top in Fig. 1 is a system consisting of a light source moved with constant velocity by a screw and an optical system giving an image of this light source over a half-reflecting mirror onto the grid at the focal plane of the telescope. From here the light goes on to the mercury mirror and back again to a point of the grid lying symmetrically to nadir with respect to the point where the light passed downwards. If the grid has different grating constants on either side of nadir similar to the principle used for collimation we have a situation similar to that shown already. The light goes on through the half-reflecting mirror and to the multiplier.

The Influence of Seeing

It is quite obvious that a star does only give information of its position when it is crossing an edge in the grid. When the star image is behind a bar or inside a slit its light does not tell how far it is from a bar, such as the light does when the image is partially obscured by an edge. This means that the position of the star is sampled discretely, namely one sample at each edge, and it is important for us to estimate the loss in accuracy compared to the ideal case of continuous sampling. In a recent paper (Høg [2]) the influence of scintillation has been evaluated and has been shown to be negligible compared to the influence of seeing.

The power spectrum of seeing measured by U. Mayer [3] in Tübingen is shown as the upper curve in Fig. 4. It corresponds to a mean square deviation of $0.''42$ for one sample. The amount of power in the frequencies below 0.2 cps is relatively unknown but is underestimated in Fig. 4.

We must consider that the star image is not a point, but a disk of finite diameter due to seeing and diffraction, and for which it therefore takes some time to cross an edge in the grid. This continuous integration at each edge modifies the power spectrum according to the formula

$$P'(f) = P(f) \frac{\sin^2 \pi f T}{(\pi f T)^2} = P(f)\,(\mathrm{dif}\,fT)^2 \tag{1}$$

where $P(f)$ is the original power spectrum as a function of the frequency f, and T is the effective integration time at each edge. An underestimation of T for an equatorial star is 0.05 sec, since

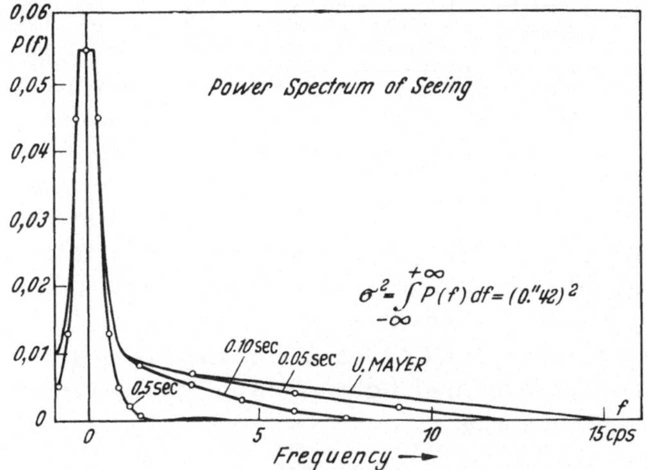

Fig. 4. Power Spectra of Seeing for several integration times

the effective diameter of the stellar disk is larger than 0.''75. The power spectra for three values of T are shown in Fig. 4 and correspond to the declinations 0°, 60°, and 84°.3.

If we further suppose that we observe each star for 20 sec and that the slits are 5'' wide we obtain for an equatorial star 60 samples since 60 edges are crossed. The mean error of the transit time for this observing time can be computed by the formula

$$\sigma^2 = \int\limits_{-\infty}^{\infty} P'(f)\,\frac{(\mathrm{dif}\,fT')^2}{(\mathrm{dif}\,f\varDelta t)^2}\,df \cong \frac{1}{T'}\sum_f P(f) \tag{2}$$

where the observing time is $T' = 20$ sec and the records are spaced at $\varDelta t$ (= 0.33 sec at the equator). The summation is over $f = 0$, $\pm 1/\varDelta t$, $\pm 2/\varDelta t$, ... cps, as indicated by the dots on the curves in Fig. 4.

The table shows first the mean error for the ideal continuous sampling during $T' = 20$ sec to be 0.''05. Our discrete sampling causes a small increase of the mean error for an equatorial star and a larger increase for a high-declination star. To the mean error contributed by the high frequencies of seeing, $f > 0.2$ cps, should however be added the contribution from the lower frequencies with periods of the order of hours and days. The experiences with the photographic zenith tubes of the US Naval Observatory suggest that these low frequencies of seeing causes a mean error of $0\rlap{.}''14$ which should be added to the mean errors in the upper

frequencies of seeing	sampling	$\delta = 0°$	$\delta = 84°.3$
$f > 0.2$ Hz	continuous	$0\rlap{.}''052$	$0\rlap{.}''052$
	discrete 5″ 30 slits / 3 slits	$0\rlap{.}''064$ 30 slits	$0\rlap{.}''097$ 3 slits
all	continuous discrete 5″	$0\rlap{.}''15$ $0\rlap{.}''15$	$0\rlap{.}''15$ $0\rlap{.}''17$

Mean error hardly encreases for discrete sampling and high declination. Observing time $T = 20$ sec

half of the table. We see from the lower half of the table that our sampling, although discrete, gives very little decrease in accuracy even for high-declination stars.

It is a pleasure for me to thank Professor SIEDENTOPF and Dr. SCHEFFLER for many useful discussions in Tübingen especially on the influence of seeing.

References

[1] Høg, E.: Astron. Abh. der Hamburger Sternw. **5**, 263 (1960). — [2] Høg, E.: Astron. J. **66**, 531 (1961). — [3] MAYER, U.: Z. Astrophysik **49**, 161 (1960).

Discussion

Scheffler: It seems to me that your computation of the accuracy of position measurement is still too unfavorable, since the power spectrum seems to go to zero for very small values of the frequency.

Høg: I should be glad if you are right, since I wanted to make a safe overestimation of the error.

Heintz: Liegen Messungen vor, aus denen man die tatsächliche Meßgenauigkeit ersieht?

Høg: Man kann etwa die Erfahrungen aus Pulkowo nehmen; bei meinem Instrument kommt noch die Integration dazu, so daß schwächere Sterne beobachtbar sein werden.

Siedentopf: As a little addition I should like to make a remark on another possibility to get an integrating micrometer for meridian circles. I had proposed this possibility a few years ago. The idea starts with the impersonal Repsold Micrometer and translates it into electronics. The proposal is to follow the star passing through the field of vision of the meridian circle with an image splitting device and give one half of the intensity to one and the other half to a second multiplier. So we can integrate the two intensities for the whole observing time without any loss, and the star gives the greatest possible information about its position during the time it passes through the field of vision. Now, the limit of the accuracy depends on the scientillation in position, since this effect determines the accuracy of the beam splitting effect. If one compares the two designs, that of Dr. HøG and our proposal, it comes out that the number of passages in the HøG proposal must be made so great that the difference between the dark and the open parts of the grid is equal to the mean amplitude of the scintillation. In this case the accuracy reached with both proposals will be the same. The beam splitting device has the advantage that it goes to a little lower magnitudes and it has the disadvantage of a greater complication in the construction. When we discussed last year the two different proposals we came to the result that for an existing meridian circle the device of Dr. HøG will be preferable. But if one is going to construct an entirely new meridian circle it may be worthwhile to consider the other proposal.

A. N. Adams (Washington): Digital Recording System and Motor Drive of the Washington Transit Circle Micrometer.

Summary: A digital data recording system has been used on the six-inch transit circle since January 1961. By means of servo repeaters and analog-to-digital converters, readings of the two micrometer screws are displayed in light banks and may be recorded by means of an accumulating printer or a tape punch. For a screw turning up to 16 RPM, the indications are read and stored to an accuracy of one milli-revolution. The store operation, controlled by the break of the standard clock, takes place in about 50 microseconds. Readings may be recorded at 1, 2, or 4 second intervals by the clock signals. A digital clock identifies the second at which the record is made. At any instant, the observer may record the readings.

A servo repeater system was designed to drive the right ascension screw of the micrometer. A ball and disc integrator with cosine input device is used as speed changer. The speed scale is nearly linear. Without discontinuity, the output speed may be varied from 16 RPM (0° Decl.) through zero RPM (90° Decl.) to −8 RPM (60° sub-polar Decl.). By means of a remotely-controlled differential synchro, the observer is able to perfect the guiding on a star. A small servo motor and control transformer are the only units of this system mounted on the telescope.

It is now possible to make observations at all declinations in the same time interval and with the same guiding technique.

The production of fundamental star catalogs by means of a transit circle involves a considerable amount of data recording and processing, both of which lend themselves very readily to modern automation techniques.

As early as 1935, the U.S. Naval Observatory instituted a program of continuing research for methods to relieve the transit circle observer of the task of recording his instrumental readings in a semi-darkened room and of the necessity of laboriously making long series of routine reductions of the most boring kind. The first marked advances in techniques in these areas resulted from the introduction of photographic methods for recording the micrometer and circle readings by C. B. WATTS [1], and the introduction of limited use of punch cards in 1943. In passing it should be remarked that at no time during these researches was it possible to inactivate the instrument for more than a few weeks at any one time. The objectives of the programs in progress and the general shortage of astronomers interested in meridian work made it impossible to do otherwise. The development and testing of apparatus had to be geared to the state of the observational program in progress. By 1955 nearly all of the calculations involved in the reductions had been transferred to the electronic computer.

Our first successful experiment with digitalization was applied to the automatic measuring engine for the circle photographs. Using a Coleman "Digitizer" and memory unit, a system capable of operating a card punch was constructed. This system has been in use since September 1956 and has proved to be very reliable, but is suitable only for a screw which is stationary when the reading is taken.

In order to digitize the readings of the micrometer, it was necessary to develop a different system capable of reading the right ascension screw "on the fly". We wished also to retain the advantages of reading the position of this screw at equal intervals of time for stars of all declinations. Such a system was designed in cooperation with engineers of the Datex Corporation of Monrovia, California, and put into operation on the Six-inch Transit Circle in January 1961 [2].

Because of the torque requirements, and weight of shaft angle encoders, and the difficulty of getting electric circuits to a transit circle micrometer, it was decided to use a synchro controlled servo system to recreate the screw readings at a remote panel rack. There the weight is no problem and the torque required to turn the encoder can be amplified by the servo system. The system developed consists of a synchro resolver, servo repeater and encoder, control chassis, and a lamp bank for each of the micrometer screws. A resolver is coupled to each of the screws by precision gears at a ratio of one to one. A similar resolver in the repeater unit produces the error signal which, when amplified, causes the motor to drive it to balance at the same phase angle as the resolver on the micrometer. The encoder is connected to the same gear train.

The encoder has two binary coded decimal plates connected by a 36 to 1 gear train. The first plate is turned at the same rate as the micrometer screw giving 1000 counts per revolution. The second plate counts revolutions up to 36. The plates are etched with a special Datex code which is able to count from 00.000 to 35.999 with only one change of brush contact or "bit" between any two successive numbers. By using a coded disk, there is no danger of losing the zero reference point as in a pulse counting system. In fact the collimation error remains as constant as with visual or photographic recording.

The control chassis contains the store-follow cricuit modules, a relay translation matrix, some of the control relays and a power supply. The store-follow modules or buffer memory are transistorized printed circuits. They are normally in the store mode. The make of the clock signal changes them to the follow mode. The clock circuit must be closed at least 50 milliseconds in order for the relays to operate properly. The break of the signal then returns the memory to the store mode in less than 50 microseconds. The positions of both screws are automatically read and stored at 1, 2 or 4 second intervals by the clock signals. The readings can also be made at any instant on command of the observer; there is, however, no time correlation in this operation. The buffer memory controls a translation matrix which makes the data available in decimal form as contact closures of relays. One set of relay contacts is connected to the lamp bank display at all times; the second set is connected to the program unit.

A digital clock serves to identify the second at which the record is stored. It is advanced by the seconds signal of the standard clock but in no way affects the read and store operation.

By means of a card reader, 10 columns of information, such as star number and right ascension may be read from the observer's prepunched setting card. The observer is also able to set up 10 digits of coded information by means of hand switches on a parameter board.

These five sets of data are brought together in the program unit where the observer may determine the format of the observational records by means of remote control switches. An adding machine and a tape punch serve as recorders.

The observer may record settings of the right ascension crew, as is required for instrumental constants. For zenith distance bisections and place of bisections, both screw readings are recorded. These operations are performed on command of the observer without time correlation. With a third switch, the observer gives a permissive signal only, which allows the standard clock to initiate the recording of the right ascension screw position and the second of time after they have been stored in their registers. At the end of each observation, a fourth switch causes a total to be taken by the adding machine and an end of message symbol to be placed in the punched tape. Although the summation of all the data recorded for each observation has no physical significance, it makes it possible for the electronic computer to check the conversion from tape records to punched cards.

At the present time, only the adding machine tapes are in use. these tapes are edited by one of the observers. For each observation he checks for irregularities in format and also writes on the tape an interpolated value of the temperature from the observer's record. The tapes are then used to produce punched cards. It is expected that these hand operations will be eliminated later this year when a tape to card converter is added to the equipment in our computing laboratory.

Simultaneously with these developments, improvements were made on the motor drive introduced by WATTS in 1933.

The new motor drive is patterned somewhat after the device developed by the Dominion Observatory, Ottawa, for the plate drive of their mirror transit circle. A ball and disc integrator, consisting of a rotating tungsten carbide disc driven by a synchronous motor, a precision-ground rotary-output cylinder, and a connecting link of two precision-steel balls, is used as a speed changer. The ball carriage is moved along a diameter of the disc to change the speed of the output cylinder. In 1961 the manufacturers added to the integrator a cosine mechanism for moving the ball carriage.

By proper choice of motor input speed and gearing, and with the cosine device set at maximum ($\cos 0° = 1$), the output speed may be made to equal that of the micrometer screw for equatorial stars. When the shaft of the cosine mechanism is turned through an angle equal to a star's declination the proper output speed is produced, including reversal for sub-polar stars.

A servo repeater system is used to transmit the output of the speed changer to the micrometer screw. It is a standard synchro system consisting of a transmitter, differential transmitter, control transformer, amplifier, and low inertia motor. The control transformer and servo motor are the only items mounted on the micrometer.

By means of the amplified error signal, the motor keeps the control transformer in phase with the control transmitter. A low inertia induction type motor is used for this purpose. They develop much more power than a synchronous motor of the same size and have a very low response time. We are able to observe at temperatures several degrees lower than before, without loss of guiding accuracy. By means of the remotely controlled differential synchro, the observer is able to perfect the guiding on a star.

It is now possible to make observations at all declinations in the same time interval and with the same guiding techniques.

References

[1] Watts, C. B.: Pub. U.S. Naval Obs. **16**, Pt. 2 (1950). — [2] Brealey, G. A.: Sky and Telescope **21**, 205 (1961).

Discussion

Fricke: Since we have just completed the 4. Fundamental Catalogue (FK4) at Heidelberg a few remarks may be permitted on the various aspects of the future of meridian circle work:

1. Any real improvement of the techniques of observation and of the reductions has to be strongly recommended, since the measurement of star positions with highest individual and systematic accuracy remains an urgent problem for the near future. Absolute and relative observations are being needed. With "relative" I mean such observations which are carried out within the system of the FK4.

2. Observations which will be carried out by absolute methods may possibly serve to improve the fundamental system. In the past, changes in the techniques of observation caused certain changes in the system of observations. In cases where consequences of these changes were not thoroughly investigated observations turned out to be of low value for the derivation of absolute proper motions. Therefore it should be obligatory that after changes in the techniques of observation one should either start with observations relative to the fundamental system or investigate the changes in the system of observations which result from the new techniques with all necessary care.

3. For general information it can be reported that the fundamental system represented by FK 4 is based on a few instruments only, in the northern sky mainly on those at Greenwich, Pulkovo and Washington, and in the southern sky mainly of Cape Observatory.

Tucker: At the Declination Bisection does the RA wire also bisect the star? If not, how far away is it (in Hour angle).

Adams: Less than $0\overset{s}{.}2$.

Walter: Everybody who does digital recording has to decide for a special code. A great facilitation would be achieved if a general agreement to a universal code would exist.

Fricke: This means a universal code should be agreed upon. The question is of importance.

Ollongren: A question was put forward by H. WALTER — Z. L. Siemens, München, regarding the presence or absence of a unified code for recording digital information of astronomical measurements on punched cards, punched tape or on some other medium. It was agreed that there is no such unified code at present, with the possible exception of the case of the ephemeris for small planets at Cincinatti (remark Strand).

I feel however that there is no real need for such strict conventions regarding the recording of information of astronomical measurements, as we have at our disposition nowadays "general pupose" computers to which the task of the decoding can be left. At the Central Computing Institute the question has been studied from a computer-oriented point of view: the conditions have been formulated to which tapes containing numerical information must comply, in order that they may be read by a general reading programme it comes out that only a few restrictions are strictly necessary, such as, for example, the condition that two numbers be separated from one another by a separate unit of information — it being allowed that this be the number's sign, and that opening (or announcing) symbols and closing symbols are necessary to mark the beginning and the end of a sequence of numbers (either the result of observation or something else) of the same kind.

K. Aa. Strand (Washington): The 61-inch Astrometric Reflector
Project of the U.S. Naval Observatory.

Summary: The U.S. Naval Observatory's 61-inch reflector to be located
at Flagstaff, Arizona, has been designed especially for astrometric work
of high precision such as stellar parallaxes.

The parallax program will be devoted to stars with apparent magnitudes
fainter than 13.5, which is the practical limit for existing refractors.

The basic features, concerning the mounting, the optics and the environ-
ments of the telescope, will be described.

Automatic features for support and alignment of the optics, for guidance
of the telescope, and read out of observational data will be discussed, including a proposed machine for rapid measurements of the photographic plates.

Introduction

High precision photographic astrometry, such as determining
trigonometric parallaxes, has primarily been carried out with
long focus refractors, because of their relatively large undistorted
fields and stable optical configurations. A notable exception is the
work of van Maanen with the 60-inch and 100-inch reflectors of
the Mount Wilson Observatory. The trigonometric parallaxes
determined with these telescopes have proven that reflectors are
suitable for high precision astrometric work when confined to
configurations of comparison stars with images close enough to the
optical axis that they are not distorted by coma.

The large number of intrinsically faint stars in the solar neigh-
borhood with apparent magnitudes fainter than 13.5 mag. are
beyond the practical limiting magnitude of even the largest
refractors now in existence as far as determination of their trigno-
metric parallaxes are concerned.

These stars, red dwarfs located at the faint end of the Main
Sequence, white dwarfs and subdwarfs, are of such great astrophysi-
cal interest that ways must be found to obtain reliable trigonometric
parallaxes for them.

The astrometric reflector of the U.S. Naval Observatory has
been designed primarily for this purpose, but will also be suited
for other types of astronomical observations.

In order to make the telescope as effective as possible, efforts
have been made to utilize the latest engineering designs incorporating automatic devices for speeding up the observational techniques
and the measurements and reductions of the observational material.

The Optical System and its Alignment

In the design of large reflectors in recent years there has been a tendency to decrease the f-ratio from the traditional $f/5$ primary to values as small as $f/3.3$ for the 200-inch Palomar telescope and $f/2.75$ for the Kitt Peak 84-inch reflector. The disadvantage of the small f-ratio for astrometric work is the small size of the coma-free field which for the 200-inch is less than 12.5 mm (Ross [1]).

The basic optical design of the 61-inch astrometric reflector consists of an $f/10$ parabolic mirror and a flat secondary mirror, 35 inches in diameter, reflecting the light through the perforation of the primary to the Cassegrain focus.

With a focal length of 15 meters, the scale is $13''.5/\mathrm{mm}$.

Using the simple theory of coma, we have

$$K'' = \frac{3}{16}\left(\frac{D}{F}\right)^2 \beta'',$$

where β is the angular distance of the star image from the axis of the mirror, D the aperture, and F the focal length of the mirror. K is the total length of the comatic image.

Assuming under good seeing conditions an optimum image size of $1''.2$ for the axial image, Ross [1] concluded from empirical data furnished by Baade that the coma would first be detectable for $K = 1''.62$ which, from the above formula, leads to a value of $\beta = 864'' = 14'.4$. The total field with no coma is, therefore, nearly $30'$ (130 mm) for the 61-inch.

Both mirrors are solid quartz, manufactured by Corning Glass Works from gaseous silica, the primary being $10\frac{1}{2}$ inches and the secondary mirror $5\frac{1}{2}$ inches thick.

The mirror support system is based upon an airtight cell housing for both mirrors.

The axial support system of the primary is provided by a layer of compressed air on the back of the mirror, with a variable air pressure, depending upon the position of the mirror. The pressure varies in range from zero, when the mirror is in a vertical position, to a pressure equal to the weight of the mirror, when in its horizontal position.

Axially the mirror is defined by three pads located in the bottom of the cell.

The radial support of the mirror is provided hydrostatically on two buoyant neoprene tubes filled with mercury. One tube surrounds the perimeter, the other the inside hole in the center of the mirror. The tubes, in addition to "floating" the mirror radially, also seal off the pressure chamber at the rear of the mirror.

The three radial defining pads are located around the hole in the center of the mirror.

The support system for the secondary mirror is similar to the system just described for the primary except that a vacuum is applied equal to the weight of the mirror when in horizontal position and zero when it is in a vertical position.

The mirror support system is provided with a pressure supply and regulating system which automatically deliver the correct pressures for both mirrors; for the primary it is a pressure of $58 \sin A$ g/cm², and for the secondary a vacuum of $30 \sin A$ g/cm² where A is the altitude of the telescope.

The optical system has a built-in collimation system which provides check on proper alignment of the two mirrors, and at the same time produces fiducial markings on the photographic plate, indicating its orientation with respect to the optical axis. Collimation will be carried out by mechanically adjusting the secondary mirror by means of motors controlled from the eye-end of the telescope. Since the secondary mirror is an optical flat, it is only required that the surface of the mirror be normal to the optical axis of the primary mirror for proper alignment. If frequent adjustments should be necessary, they can be made by means of an automatic system incorporating a radiation tracking transducer.

The Mounting and the Support of the Telescope

The telescope mounting belongs to the symmetrical class of fork and yoke design, and resembles in many respects the 120-inch Lick telescope with its unequal armed parallelogram truss structure for the tube.

The distances from the center of the declination axis to the mirror cells of the primary and secondary mirrors are 1.8 and 5.8 meters respectively, using a nearly one-fourth-point balance position.

The total weight about the declination axis is 14740 kgs, and the primary and secondary mirrors with their support systems weigh 5000 and 750 kgs respectively.

Since only the Cassegrain focus is being used, it is essential that enough working space is left between the mirror cell and the fork to allow the observer access in all positions of the telescope. It is also essential that the stiffness of the fork be preserved by not making it unduly long.

The fork is a welded structure of 2.5 cm steel plates reinforced by internal stiffening partitions.

The total length of the fork from the center of the declination axis to the face of the fork bearing is 4.7 meters, allowing a working space of 1.2 meters between the focal plane and fork when the telescope is directed towards the North Pole.

The rotating weight of the telescope about the polar axis is 36300 kgs and is carried by a combination of oil pad and standard ball bearings.

The north-polar bearing consists of three oil pads supporting most of the weight of the telescope, both radially and axially. Each pad consists of a cylindrical surface, matching the journal surface with a recessed control cavity with a diameter of half that of the pad. The oil under high pressure flows into the cavity and is allowed to flow out to the collection channel around the edge of the pad. The oil film between the journal and the surface of the pad ranges in thickness from 75 to 125 microns.

Special needle valves, one for each pad, allow adjustment of the oil film thickness for the particular temperature conditions prevailing during a normal night's observation period.

The right ascension drive of the telescope consists of a single worm wheel used for both slewing and tracking.

Corrections to the guiding of the telescope will be made directly upon the right ascension and declination drives by means of 1/70 horsepower guide motors. The guiding will be carried out automatically by means of a photoelectric guider of the Weitbrecht type [2], which uses a pyramid-shaped prism and four 1 P21 photomultiplier tubes as position indicators. The guider used is part of the instrument mount and the prism is capable of scanning over a 10×25 cm $(25' \times 60')$ area of the field of the telescope for the purpose of selecting a suitable guide star.

The camera, which has been designed for the parallax observations, is provided with automatic timing and plate transport. Exposure times can be selected from 10 seconds to one hour, and by means of selector switches any or all of the six positions of the plate can be selected for exposures in a sequence which is carried out automatically. A separate switch has been provided to permit the taking of conventional single exposures without any movement of the plate.

The automatic features of the guiding system and the camera should provide maximum utilization of the observing time.

The Site of the Telescope

The astrometric reflector will be housed in a new observatory building at the Flagstaff Station of the U.S. Naval Observatory, approximately five miles west of the city of Flagstaff, Arizona.

The station is located 2300 meters above sea level and has excellent observing conditions in regard to the number of clear nights, transparency, and seeing, as experienced from the operation of the 40-inch Ritchey-Chrétien telescope during the past six years.

The observatory building has been designed with the principal aim of minimizing the deleterious effects of the higher daytime temperatures upon the nighttime operation of the telescope.

For this reason the dome has double walls with air space and insulation in between. Similarly, the stationary structure of reinforced concrete below the dome is protected against heating up during the daytime by cellular glass insulation and an aluminum shield spaced 6 inches from the wall to allow free flow of air.

Measurements and Reduction of the Photographic Plates

It is expected that the efficiency of the telescope combined with the favorable observing conditions will produce such a large amount of plate material that up-to-date automatic means for measuring the plates will be necessary.

Studies are currently being made of a system which, from input data in the form of punched cards, would automatically measure the positions of the selected star images on the photographic plates in a prescribed sequence recording the output data on punched cards or tape.

The input data would be obtained from the initial plate of a series, which would be measured manually.

The planned machine should be capable of measuring with the same precision as is attained with the conventional comparators now in use, while at the same time speeding up the measuring process.

Availability of Components for Automatic Comparators

All of the required components for an automatic measuring machine are available in one form or the other at the present time; therefore, it is primarily a problem of selecting the units which will provide the simplest and most reliable system.

Besides the conventional screw-type comparator, there are such other measuring devices as the precision metal belts, linear scales (quartz, glass or steel) in combination with a micrometer, fringe counting micrometers, and Farrand Inductosyn linear micro-positioning systems.

There are also a number of different image-centering devices which would be adaptable for such a machine.

Once the measured data are on punched cards or tape, the reduction of the data and the derivation of the results become a simple matter with present day high speed computers.

References

[1] Ross, F. E.: Astrophys. J. **81**, 156 (1935). — [2] Weitbrecht, R.H.: Rev. Sci. Instr. **28**, No. 2, 122 (1957).

Discussion

Köhler: Da nur ein einfacher Parabolspiegel verwendet wird, ist am Bildfeldrand (29′) Koma und Astigmatismus schon merklich. Die Bilddefinition am Bildrand muß doch schon merklich schlechter sein als in der Bildmitte. Warum verwenden Sie kein afokales Korrektionssystem?

Strand: It is only intended to use the field within 15′ from the center of the optical axis for high precision work. At the edge of this field the effect of coma is still negligible. While a zero corrector has been incorporated in the optical system it will only be used for survey fields 1° in diameter, but not for high precision astrometric work.

Schwesinger: Why is it proposed to make the mercury tubes in the mirror support system of teflon?

Strand: Teflon is not chemically affected by mercury. However, it may not be sufficiently elastic, and for this reason a material such as neoprene (See text) might be used. The sealing of a mercury system is always a problem.

W. Schuler (Neuchâtel): Das vollautomatische photographische Zenit-Teleskop und der numerierende Chronograph des Observatoriums von Neuenburg. (Mit 7 Textabbildungen.)

Zusammenfassung: 1. Die Beobachtungsmethode am photographischen Zenit-Teleskop prädestiniert dieses Instrument geradezu für eine Vollautomatisierung. Ein elektromechanisches Programmgerät löst zu den gewünschten Startzeiten die zur Registrierung der Sterndurchgänge notwendigen Funktionen aus. Die Informationen werden dem Gerät mittels gelochtem Fernschreiberband zugeführt. Da in Sternzeit programmiert wird, kann der Programmstreifen einer Nacht ungefähr einen Monat unverändert verwendet werden. Eine besondere Anpassung des Instrumentes erlaubt die Anzahl der Sternregistrierungen zu verdoppeln, wodurch die wartungsfreie Beobachtungsdauer auf 8 Std gebracht wird. Ein Regen-Detektor schützt das Instrument vor plötzlichem Witterungsumschlag.

2. Der summierende Chronograph bildet automatisch das Zeitmittel einer bestimmten Anzahl von Impulsen (z. B. Mikrometerkontakte von astronomischen Instrumenten).

Die angewandte Integrationsmethode basiert auf Impulszählung von veränderlicher Frequenz. Am Observatorium von Neuchâtel wird der Prototyp dieses Chronographs bei den Beobachtungen am Astrolab Danjon verwendet. Er druckt unmittelbar nach jedem Sterndurchgang seine gemittelte Durchgangszeit. Die kommerzielle Herstellung des Gerätes wird vom Département Oscilloquartz der Firma Ebauches S.A., Neuchâtel, unternommen.

Das photographische Zenit-Teleskop eignet sich so vorzüglich für einen vollautomatischen Betrieb, daß man leicht auf den irrigen Gedanken verfallen könnte, die Beobachtungsmethode sei eigens im Hinblick auf die Automatisierung entwickelt worden. Der überaus einfache Aufbau des Instrumentes steht aber in keinem ursächlichen Zusammenhang mit der Automatisierung, welche ursprünglich nur so weit durchgeführt wurde, als es die Beobachtungsmethode überhaupt erforderte. Bei diesem Minimalaufwand handelt es sich darum, die photographische Platte mit konstanter Geschwindigkeit dem Stern nachzuführen, während des Sterndurchgangs in genauen Intervallen vier Belichtungen aufzunehmen, sowie zwischen den einzelnen Belichtungen das Rotary eine halbe Drehung um die vertikale Achse ausführen zu lassen (Rotary ist die Bezeichnung für den oberen, drehbaren Teil des Rohres, auf welchem das Objektiv und der Plattenwagen mit seinem Mikrometerantrieb montiert sind).

Mit der automatischen Ausführung der erwähnten Operationen ist jedoch nur ein erster Schritt zur Automatisierung gegeben, der wenigstens eine Fernsteuerung ermöglicht. Man benötigt nun immer noch einen Operateur, der den Plattenwechsel vornimmt und jeweils zur gegebenen Zeit auf den Knopf drückt, um die Sternregistrierung zu starten.

Solange eine menschliche Überwachung ohnehin vorhanden sein muß, ist man geneigt, das Starten von Hand in Kauf zu nehmen.

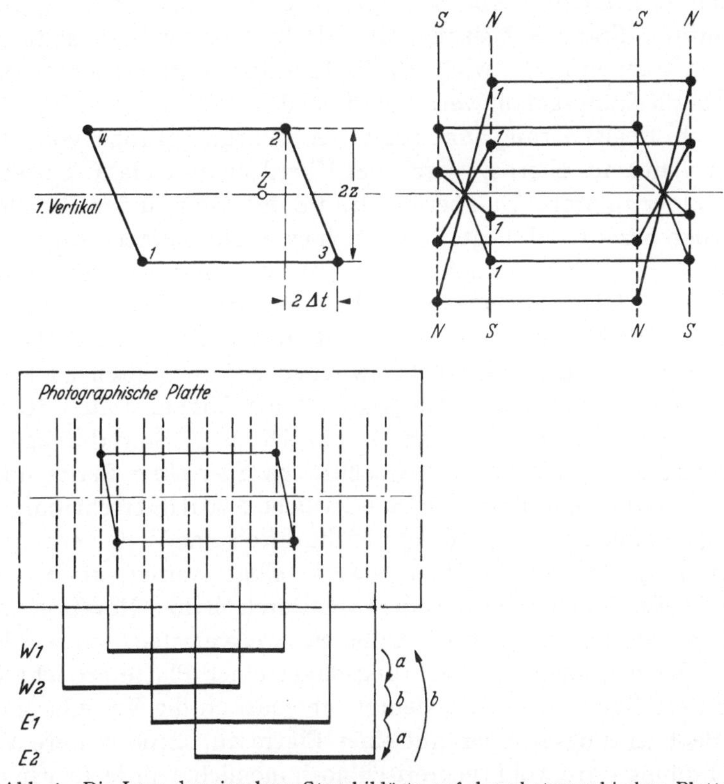

Abb. 1. Die Lage der einzelnen Sternbildäste auf der photographischen Platte
(Erläuterungen siehe Text)

Wenn aber einmal der alle 2 Std notwendige Plattenwechsel automatisch erfolgt, ist die Beanspruchung eines Operateurs nur für das Starten der Aufnahmen nicht mehr zu rechtfertigen.

Beim Instrument von Neuchâtel kann leider der automatische Plattenwechsel nicht ohne tiefgreifende Konstruktionsänderungen verwirklicht werden und man mußte sich daher mit einer besseren Flächenausnutzung auf der einen Platte behelfen.

Abb. 1 zeigt das Zustandekommen der einzelnen Sternbildäste auf der Platte. Jeder Sterndurchgang ergibt ein Parallelogramm, dessen Schiefe von der Genauigkeit des Startes abhängt und dessen Höhe durch die doppelte Zenitdistanz des Sternes im Meridian gegeben ist. Die einzelnen Parallelogramme überlagern sich und unterscheiden sich nur durch die Verschiedenheit der Zenitdistanzen der einzelnen Sterne, welche denn auch ihre Identifizierung ermöglicht. Abb. 1 zeigt auch, wie man mittels einer konstanten Startverzögerung, die nördlichen von den südlichen Sternen trennen kann, wodurch die Identifizierung erleichtert und die Überdeckungsgefahr vermindert wird.

Nach Registrierung von einer Beobachtungsgruppe von etwa einem Dutzend Sterne, wird die Überdeckungsgefahr einzelner Sternbilder zu groß. Da aber das Bild Z des Zenit zu den Parallelogrammen exzentrisch liegt, ergibt sich eine sehr einfache Ausweichmöglichkeit. Wird nämlich die Ausgangsstellung des Rotary und damit der Platte um eine halbe Drehung geändert, so fallen die neuen Parallelogramme neben die alten, wobei einer der beiden neuen Äste genau in die Mitte zwischen die zwei alten zu liegen kommt. In dieser neuen Ausgangsstellung können nun während zwei weiteren Stunden nochmals ein Dutzend Sterndurchgänge registriert werden. Um die Beobachtungsdauer einer Platte nochmals zu verdoppeln, muß nun eine seitliche Plattenverschiebung zu Hilfe genommen werden, die sehr leicht durch eine Veränderung der Nullstellung des Mikrometers bewerkstelligt werden kann. Auf diese Weise können wiederum zwei durch halbe Rotation verschobene Sterngruppen in die früheren zwei verschachtelt werden, womit die maximale Beobachtungsdauer einer Platte erreicht ist. Die untere Skizze in Abb. 1 deutet schematisch die Verteilung der vier Beobachtungsgruppen auf einer Platte an. Eine weitere Verschachtelung wäre wohl theoretisch noch möglich, würde aber in der Praxis sehr oft zu Identifizierungsschwierigkeiten führen.

Die zwei zusätzlichen Operationen: halbe Drehung des Rotary und Verstellen des Mikrometernullpunktes sind sehr leicht zu automatisieren. Die halbe Drehung ist eine Teilbewegung des Programmes für die 4 Belichtungen pro Sterndurchgang und das Verstellen des Mikrometers bedingt lediglich ein zeitlich genau begrenztes Einkuppeln des Mikrometergetriebes auf den zur Nachführung verwendeten Synchronmotor.

Das den Operateur ersetzende Steuergerät wurde auf folgende Weise verwirklicht: Ein Zählwerk zählt die Sekundenimpulse einer Sternzeituhr und zeigt den momentanen Uhrstand in Sternzeitsekunden an. Es ist gekoppelt mit einem Speicherzählwerk das als Information die Codenummer des auszuführenden Befehls, sowie die Startzeit in Sternzeitsekunden enthält. Erreicht das Sekundenzählwerk den im Speicher registrierten Uhrstand, so wird durch eine Koinzidenzschaltung der vorgemerkte Befehl ausgelöst. Gleich-

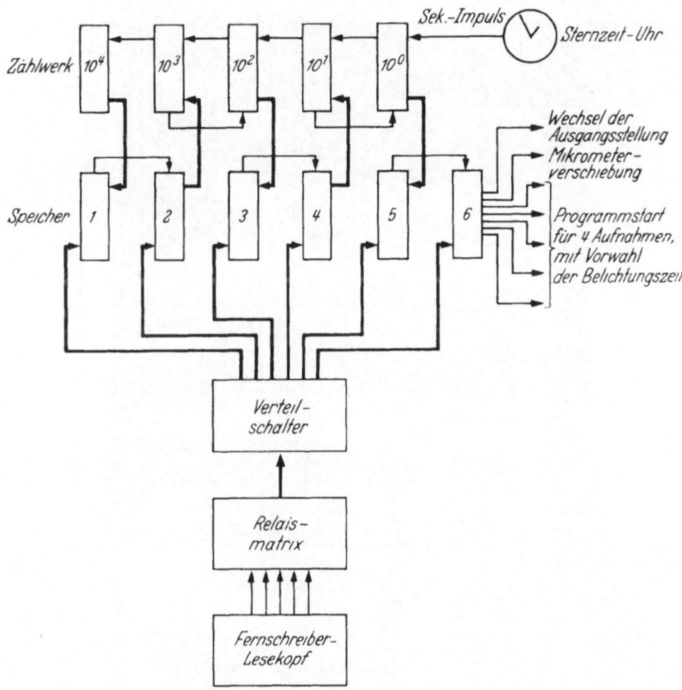

Abb. 2. Prinzipschema des Steuergerätes

zeitig wird dem Speicher eine neue Information mittels Fernschreiberlochstreifen zugeführt.

Das Sekundenzählwerk besteht aus einer Kette von Telephonschrittschaltern die von Stufe 10 automatisch wieder auf Stufe 0 schalten, wobei der nächsthöhere Schalter um eine Stufe vorrückt. Bei Erreichen der Zahl 86400 schaltet die ganze Kette innerhalb einer Sekunde auf null.

Das Speicherwerk ist ähnlich wie das Zählwerk aufgebaut, besitzt jedoch einen Schrittschalter mehr für die Codenummer des auszuführenden Befehls. Der Fernschreiberlesekopf tastet die fünf Lochpisten gleichzeitig ab, und betätigt eine Matrix von 5 Relais zu je 10 Kontakten. Der Ausgang der Matrix wird von 10 Leitungen gebildet, wovon diejenige unter Spannung gesetzt wird, welche der vom Streifen abgelesenen Ziffer entspricht. Die Relaismatrix kann nur die Ziffern 0 bis 9 übersetzen. Sie ignoriert jedes andere Zeichen des Telexcodes.

Abb. 3. Das Steuergerät

Die zehn Matrixausgänge führen auf einen Verteiler, der die erste aufgenommene Ziffer auf den ersten Zähler des Speichers gibt und diese entsprechend einstellt. Die nächste Ziffer kommt auf den zweiten Zähler usw. Die letzte Ziffer auf dem 6. Zähler bestimmt die Art des auszuführenden Befehls (Abb. 2).

Im Moment der Auslösung des Befehls schaltet der Verteiler-schalter auf Null, der Lochstreifen rückt vor und die nächsten 6 Ziffern werden gespeichert. Um das Instrument bei eventuellen Störungen vor Schaden zu bewahren, wurden auch verschiedene Sicherheitsschaltungen eingebaut, die das Instrument gegebenen-falls stillegen.

Auch mußte man sich gegen plötzliche Witterungsumschläge vorsehen. Um das Dach der Beobachtungshütte mit möglichst kleinem Aufwand automatisch schließen zu können, wurde über dem Instrument nur ein Lichtschacht von $1\ m^2$ Querschnitt offen-gelassen, der mittels eines Klappdeckels hermetisch geschlossen werden kann.

Bei einsetzendem Regen oder Schneefall, löst ein Detektor die Arretierung des Deckels, der von selbst zuklappt. Ebenso wird das Dach auch bei Programmende vor der Morgendämmerung auf den Befehl „Hauptschalter aus" geschlossen. Die belichtete Platte ist somit vor Tageslicht geschützt. Die Herstellung und Änderung des Programmstreifens erfolgt sehr leicht auf einem mit Streifenlocher ausgerüsteten Fernschreiber. Das gesamte Beobachtungsprogramm von 118 Sternen findet auf einem Streifen von etwa $2\ m$ Länge Platz. Da die Startzeiten nur auf 1—2 sec genau erfolgen müssen, genügt im allgemeinen eine monatliche Programmänderung, die der Präzession Rechnung trägt.

Das auf Abb. 3 wiedergegebene Steuergerät wurde 1959 in Betrieb genommen. Es hat sich seither sehr gut bewährt, denn es gestattet ohne jeglichen Personalaufwand jede Aufheiterung auszu-nützen.

Eine französische Beschreibung des Gerätes ist 1959 erschie-nen [1].

Der summierende Chronograph

wurde von Herrn Prof. J. P. BLASER, dem früheren Direktor des Observatoriums in Zusammenarbeit mit Herrn CHARLES WYSER konstruiert. Eine detaillierte Beschreibung wurde ebenfalls 1959 veröffentlicht [2].

Der Chronograph wurde für die Registrierung der Beobach-tungen am Astrolab Danjon gebaut mit dem Ziel, den zeitlichen Mittelwert der Mikrometerkontakte zu bilden und als Resultat jedes Sterndurchganges zu drucken.

Die verwendete Methode der Mittelwertbildung basiert auf dem Auszählen der Periodenzahl einer von der Mikrometerstellung x

abhängigen Frequenz. Bei linearer Abhängigkeit der Frequenz
$f(x) = a\,x + b$, kann man leicht zeigen, daß die Summe N der
Perioden gleich dem gesuchten Mittelwert $\bar{\iota}$ wird, sofern $a = -1/x_n$
und $b = 1$ mit $x_n = $ Endkoordinate des Mikrometers, also für:

$$f = 1 - \frac{x}{x_n}.$$

Da wir jedoch am Astrolab wie auch bei anderen Beobachtungs-
instrumenten diskontinuierliche Messungen mit einzelnen Mikro-

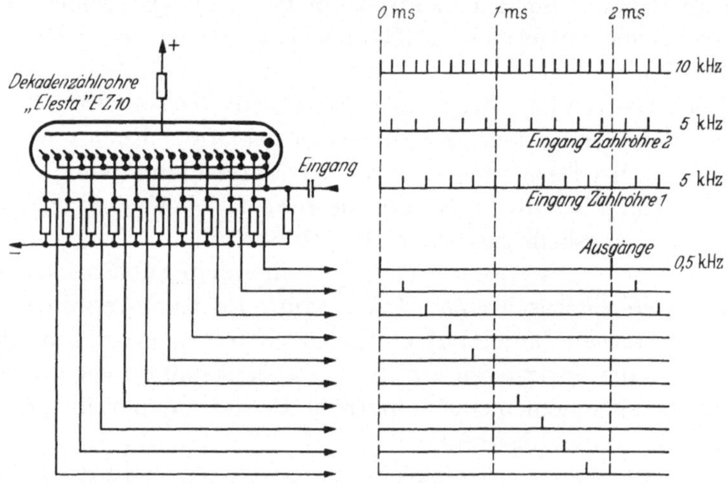

Abb. 4. Die Impulsverteilung an den Kathoden einer Dekadenzählröhre

meterkontakten machen, muß mit einer sprunghaft veränderlichen
Frequenz gearbeitet werden.

Für 20 äquidistante Kontakte ersetzen wir in der Frequenz-
formel x_n durch 20 und x durch i, mit $i = 1, 2, 3, \ldots, 20$.

Für das Intervall zwischen den Kontakten i und $i + 1$ ergibt
sich dann die Frequenz

$$f_{i \to i+1} = 1 - \frac{i}{20}$$

und mit der Konstante $k = 20\,f_0$ erweitert:

$$k\,f_{i \to i+1} = f_0\,(20 - i).$$

Die Periodenzahl N wird dadurch ebenfalls um den Faktor K
heraufgesetzt und wir werden als Mittelwert nun $\bar{\iota} = N/k$ haben.

Die Bestätigung ergibt sich ohne weiteres beim Auszählen von N nach obiger Formel

$$N = 20\,f_0\,t_1 + 19\,f_0\,(t_2 - t_1) + 18\,f_0\,(t_3 - t_2) + \cdots + 2\,f_0\,(t_{19} - t_{18}) +$$
$$+ f_0\,(t_{20} - t_{19}) = f_0\,(t_1 + t_2 + t_3 + \cdots + t_{20}) = 20\,f_0\,\bar{t}.$$

Es handelt sich also darum, eine gegebene Frequenz $20f_0$ bei jedem Mikrometerkontakt um den konstanten Betrag f_0 herabzusetzen. Nach dem 20. Kontakt fällt die Frequenz auf Null, die Zählung kommt zum Stillstand und der erreichte Wert braucht nur noch registriert zu werden.

Als Zählgerät wurde ein elektronischer Zähler Berkeley mit seinem Zifferndrucker gewählt. Die variable Frequenz setzt sich aus zwanzig unabhängigen Impulsreihen zusammen, deren Zustandekommen in Abb. 4 und 5 veranschaulicht ist.

Wird am Eingang der Dekadenzählröhre eine Frequenz von 5 kHz angelegt, so erhält man am Ausgang jeder Hauptkathode eine Frequenz von 500 Hz. Jeder Impuls einer Kathode ist um 0,2 ms zu demjenigen der benachbarten Kathode verschoben. Führt man die 10 Ausgänge zusammen, so erhält man wieder die Eingangsfrequenz von 5 kHz. Jedesmal, wenn ein Kathodenausgang unterbrochen wird, sinkt diese Frequenz um 500 Hz.

Für die 20 benötigten Frequenzsprünge gehen wir von einer Bezugsfrequenz von 10 kHz aus, deren Impulse durch eine Flipflop-Schaltung alternierend auf zwei Zählröhren verteilt werden (Abb. 5).

Die 20 Hauptkathoden sind über eine Dioden-Unterbrecherschaltung vereinigt. Die an den Kathoden auftretenden rechteckigen Impulse (A) erscheinen an den Dioden differenziert (B). Die positiven Impulse werden von den Dioden blockiert, negative Impulse passieren die Dioden, solange die Punkte K_i auf Null-Potential sind. Erscheint aber auf einem K_i eine positive Polarisation, so werden an der entsprechenden Diode auch die negativen Impulse blockiert. So erhält man am Ausgang negative Impulsreihen, deren effektive Frequenz von 0 bis 10 kHz um Sprünge von 500 zu 500 Hz variierbar ist.

Die positiven Sperrspannungen werden von einer Zählkette geliefert, bestehend aus 24 Kaltkathoden-Röhren (Cerberus GR 20).

Die modifizierten Mikrometerkontakte des Astrolabs werden den Startern zugeführt, welche so polarisiert sind, daß bei jedem Mikrometerkontakt eine weitere Röhre gezündet wird. Beim 24. Kontakt betätigt der Anodenstrom der letzten Röhre ein Relais, das nach einer gewissen Verzögerung die Anodenspannung unterbricht und damit die ganze Kette löscht. Die 20 Dioden sind auf die Kathoden

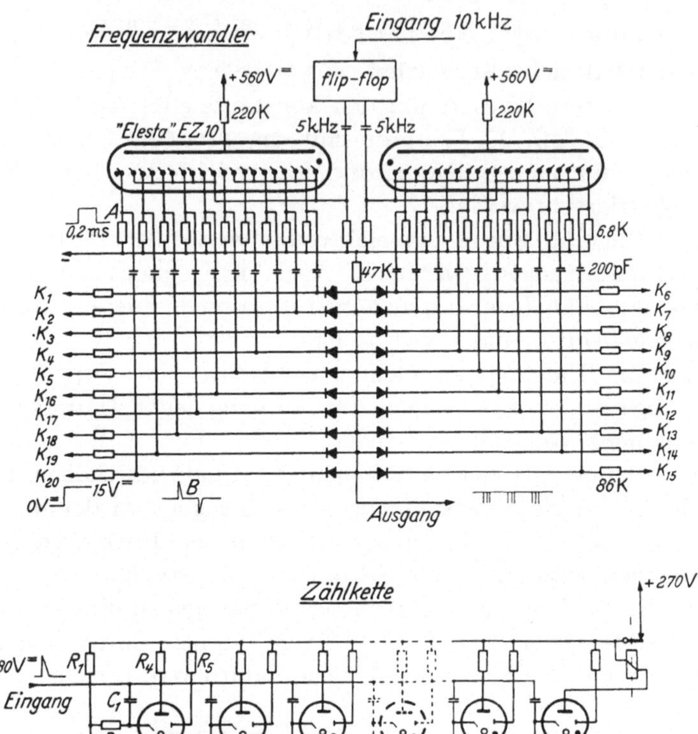

Abb. 5. Schaltschemata von Frequenzwandler und Zählkette

der Röhren 3 bis 22 geschaltet. Sobald eine Röhre durch den entsprechenden Mikrometerkontakt gezündet wird, entsteht am Widerstand R 7 ein Spannungsabfall, der die Diode positiv polarisiert und somit eine Frequenzverminderung von 500 Hz bewirkt.

Die ersten und letzten zwei Mikrometerkontakte werden in die Mittelwertbildung nicht einbezogen. Abb. 6 gibt das vollständige

Funktionsschema des Chronographs. Die Frequenz 10 kHz wird von einem Quarzoszillator geliefert. Solange kein Stern beobachtet wird, wiederholt sich in einem Zyklus von 20 sec eine durch die Steuerscheibe C geregelte Synchronisierung mit der Referenzuhr. Der Kontakt C schließt sich bei Sekunde null, 20 und 40 während $^3/_4$ sec. Er speist das Relais A, welches den Zählerstand auf dem „Berkeley" löscht und gleichzeitig eine Registrierung auf dem

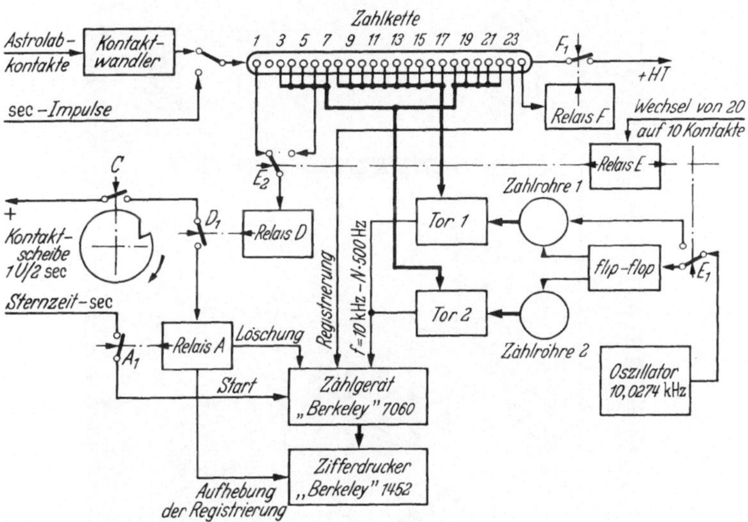

Abb. 6. Prinzipschema des summierenden Chronographen

Druckschreiber unterbindet. Ein weiterer Kontakt A_1 des Relais A bildet ein Tor für den Sekundenkontakt 0, 20 oder 40, welcher den Zähler Berkeley von neuem startet. Bei einem Sterndurchgang wird durch den ersten Mikrometerkontakt die Röhre 1 der Zählkette gezündet und damit das Relais D unter Strom gesetzt. Sein Kontakt D_1 öffnet den Stromkreis von A und hebt dadurch den Synchronisierungszyklus für die Dauer des Sterndurchgangs auf. Bei jedem Kontakt zündet eine weitere Röhre. Wenn Röhre 22 zündet, fällt die Frequenz von 500 Hz auf null. Die Registrierung des erreichten Zählerstandes wird durch Röhre 23 ausgelöst und Röhre 24 löscht die ganze Zählerkette durch Unterbrechung des Kontaktes F_1. Beim nächsten Synchronisierungszyklus startet der Sekundenkontakt wiederum den Zähler, der von neuem die Initialfrequenz 10 kHz auszählt.

Der summierende Chronograph druckt also direkt auf die Zehntausendstel-Sekunde die gemittelte Durchgangszeit eines Sternes mit einer Genauigkeit von 0,13 ms, ausgehend von einem Uhrstand der ein zunächst unbekanntes Vielfaches von 20 sec darstellt. Der

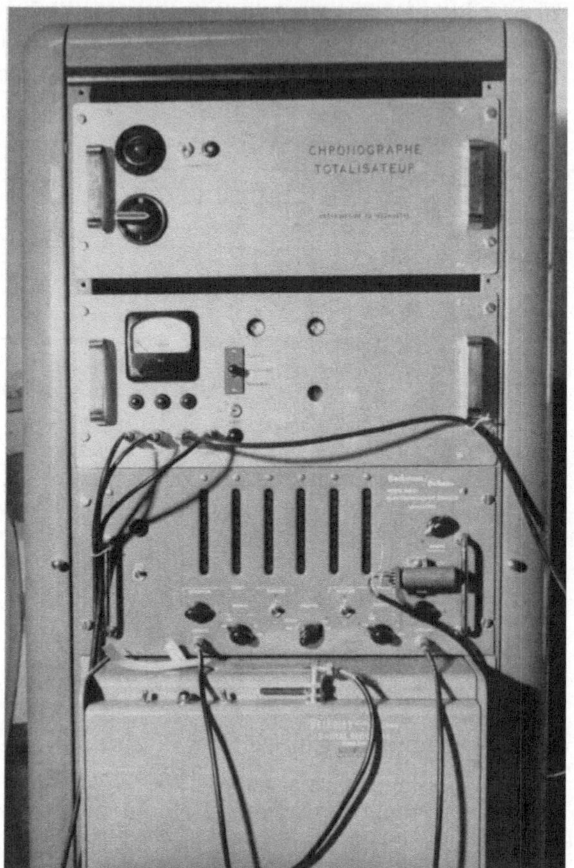

Abb. 7. Der summierende Chronograph des Observatoriums von Neuchâtel

volle Stand der Durchgangszeiten in Stunden und Minuten ist bei Bedarf leicht zu ergänzen. Er könnte auch durch Einbau einer Addiermaschine gewährleistet werden. Der am Observatorium gebaute Prototyp, der in Abb. 7 wiedergegeben ist, arbeitet nun schon seit 3 Jahren ohne nennenswerte Störung. Auf Grund der gemachten Erfahrungen wird gegenwärtig von der Firma Ebauches S.A., Département Oscilloquartz, ein volltransistorisiertes Modell auf kommerzieller Basis entwickelt.

Literatur

[1] WYSER, CH., J.-P. BLASER et W. SCHULER: L'automatisation de la lunette zénithale photographique. Publ. Observatoire Neuchâtel No. 7 (août 1959). — [2] BLASER, J.-P., et CH. WYSER: Chronographe totalisateur pour instruments astronomiques. Publ. Observatoire Neuchâtel No. 4 (août 1959).

Diskussion

Høg: Sollte man nicht besser die Sterndurchgänge am PZT photoelektrisch registrieren?

Schuler: Das ist ein rein technisches Problem. Da die Fokalebene in die zweite Hauptebene des Objektives, die höchstens 1—2 cm unter dem letzten Linsenscheitel liegt, reflektiert werden muß, fehlt der Raum für die zur photoelektrischen Registrierung notwendige Apparatur. Eine Bohrung im Objektiv oder eine Bündelablenkung könnte zu einer Lösung führen. Es fragt sich dann aber, ob die beim heutigen PZT sehr weit getriebene Unabhängigkeit von Instrumentalfehlern nicht teilweise eingebüßt werden muß.

R. H. Tucker (Herstmonceux): EPIC: An Electronic Punching
and Indicating Chronograph. (With 1 Figure.)

Summary: After a brief survey of the history of timing methods in
astronomy, it is pointed out that at the present time an accuracy of one
millisecond is both desirable and attainable by various methods.

The policy adopted at R.G.O. Herstmonceux, is that data should be
obtained in machine-readable form at the earliest possible stage. EPIC was
developed in accordance with this policy to give clock times of contact
signals directly on punched cards.

EPIC is then described in general terms, and a more detailed explanation
of the process of operation is given, followed by illustrative numerical
examples.

The punched cards produced by EPIC are processed daily by computer,
and these preliminary reductions are examined so that suspect observations
may be investigated without delay.

A recent addition to the equipment is a control panel which enables
the observer to record descriptive data on the punched cards at the time
of observation, and so avoiding the need for hand-punching at a later stage.

Timing Methods for Transit Observations

The earliest transit telescope observations were made by the
"eye and ear" method, in which the times of passage of the star
image across fixed wires were estimated by an observer listening
to the seconds beats of a clock. The estimated times were generally
consistent to $\frac{1}{10}$ second, but subject to rather large systematic
errors depending on the observer and the speed of the star. The
ease and rapidity of the observations were later improved by providing the observer with a hand tapper in electrical connection
with a recording chronograph. The estimation of the time was
postponed until the chronograph records of the clock and hand
tapper were read off at leisure.

The next advance was the introduction of Repsold's "impersonal" micrometer in which a wire was moved by hand, and
maintained in coincidence with the star image. The micrometer
screw carried a contact drum which sent signals to be recorded on
the chronograph. The personal errors were reduced to less than
$\frac{1}{10}$ second with this apparatus, and a further reduction was
achieved when the wire and eyepiece were moved by a motor,
leaving the observer only the task of maintaining an accurate
bisection of an apparently stationary image.

Great progress has been made in astronomical timekeeping with
the installation of quartz-crystal clocks in many observatories.

Seconds signals accurate to $\frac{1}{1000}$ second are available without difficulty, and it becomes necessary to plan a similar precision in the chronographic recording process in order to preserve the inherent accuracy of the micrometer contact signals and the clock pulses.

It is possible to approach millisecond accuracy by using various modern types of pen or stylus chronographs, and direct-printing chronographs have been developed which give an immediate

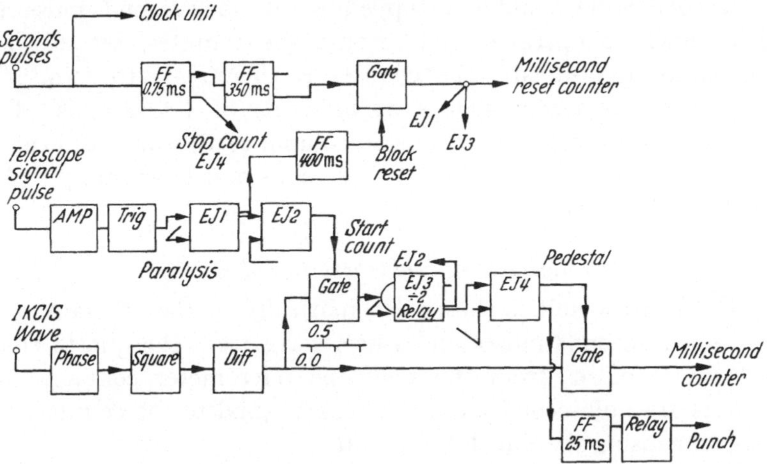

Fig. 1. Schematic diagram of the control unit of EPIC

printed record to $0^s.001$. Practicable techniques have also evolved which depend on the photography of a rotating disc or micrometer head by the light of an accurately timed flash.

All these methods require further processes to be performed before the data can be presented to an automatic computer for calculation, and the equipment now to be described was developed to overcome this difficulty by the immediate production of data in machine-readable form.

EPIC — General Description

The equipment comprises a clock unit, a millisecond counter unit and a control unit mounted in a standard rack, together with an automatic card punch (I.C.T. type 029).

The clock unit is operated without interruption by the sidereal seconds pulses from the observatory's standard quartz clock, and is set one second slow on local sidereal time. The millisecond counter is started by the arrival of a telescope micrometer contact signal, and stopped at the next following clock second. The clock, being one second slow, now indicates the time at the beginning of the second in which the signal arrived, and the millisecond counter shows the complement to 999 of the decimal of second of the signal. The millisecond counter is connected to the punch so that 0 is punched as 9, 1 is punches as 8, and so on, with the result that the decimal of the second is correctly punched.

As soon as the counter is stopped by the clock seconds pulse, the card punch is operated and records the minutes, seconds and decimal of the signal. The counter is then reset to zero. The apparatus can accept signals at intervals of 1.4 seconds. Five signal times are recorded on one 80-column card, which also carries indicative data of various kinds. The days and hours are punched for the fifth tap only.

Control Unit and Operation of EPIC

The control unit is shown schematically in the diagram. The inputs are seconds pulses and 1 kc/s wave from the quartz clock, and signal pulses from the telescope micrometer contact. The outputs are millisecond pulses and reset signal to the counter, and a signal causing the punch to operate.

The millisecond pulses are generated from the 1 kc/s wave. A short negative pulse occurs every millisecond on the millisecond (0.0) and a short positive pulse every millisecond on the half-millisecond (0.5). The input wave is adjusted in phase so that the negative (0.0) pulse at the full second coincides with the standard seconds pulse. The arrival of a signal from the telescope operates the input circuit and renders it insensitive to further signals. The first positive (0.5) pulse after a telescope signal is admitted to a divide-by-two relay which is reset by the next pulse (1.5). On being reset, the relay blocks the ensuing positive pulses (2.5, 3.5 etc.) and admits the negative (0.0) pulses to the millisecond counter. The count continues until terminated by a slightly delayed clock pulse at 0.75 milliseconds past the next full second, and punching then begins. The counter is reset to zero at 350 milliseconds past the full second by a delayed clock pulse, which also restores the signal input circuit in readiness for the next telescope signal.

Precautionary reset is applied to the counter, signal input circuit and divide-by-two relay at every second to minimise the risk of spurious operation by stray signals. Reset is blocked for 400 milliseconds after the receipt of a telescope signal, to ensure that reset of the counter does not occur while a count is actually in progress.

The complex operation sequence is designed to avoid pulse-coincidence difficulties, to give a decimal correctly rounded-off and in correct relation with the integral second, and to obviate the need for close tolerances in phase and lag.

Numerical Examples

Signal pulse at 583.2 milliseconds after full second.
Divide-by-two relay operates 583.5—584.5

    Pulses to counter 585 586 587 1000
    Count   001 002 003 416
999 — complement punched   583

Signal pulse at 246.8. Reset at 350.0 blocked.
Divide-by-two relay 247.5—248.5

    Pulses to counter 249 250 251 1000
    Count   001 002 003 752
    Punch   247

Signal pulse at 999.4. Reset blocked at 1350.0
Divide-by-two relay 999.5—1000.5

    Pulses to counter NIL (1000.5—1000.75)
    Count   000
    Punch   999 (OLD SECOND)

Signal pulse at 999.6. Reset blocked at 1350.0
Divide-by-two relay 1000.5—1001.5

    Pulses to counter 1002 1003 1004 2000
    Count   001 002 003 999
    Punch   000
             (NEW SECOND)

Preliminary Examination of Recorded Transits

The immediate availability of the transits on punched cards has made it possible to perform the preliminary examination of the recorded transits without delay. After each night's work the cards are taken to the I.C.T. 1201 computer of the N.A.O. at Herstmonceux and by 10 a.m. the results are sent to the Meridian Department. The computer first lists the data punched on each card and then prints the four first differences between the five recorded times. Finally, it gives the mean of the five times, the mean of the four differences and a computed time directly comparable with the time given in the observing catalogue. This facilitates verification that the correct star has been observed.

The computer also gives a numerical indication of the scatter of the five times from linearity, expressed in arc, and prints an alarm signal if any of the four differences departs by more than 1 part in 30 from the mean of the four.

The prompt delivery of these results makes it possible to examine suspect observations and to make a decision on rejection before marking off the stars in the observing list for the next night's observers.

Punching of Descriptive Data by Observer

A recent innovation at Herstmonceux is the installation of a control panel from which the observer or his assistant can punch a report of the stars observed. After each transit has been punched automatically by EPIC, the observer sets up the star number and other information on push-buttons, checks his setting by illuminated indicators, and presses a button which causes the 029 punch to transfer the information to a fresh card.

The pack sent for computer processing now includes report cards interspersed with the transit cards. All the cards are listed together, presenting the information conveniently for examination. The great advantage of the innovation is that the data is produced in machine-readable form, thus saving the process of hand-punching from manuscript, and the corresponding checking stage.

A. Behr (Göttingen): Design of an Automatic Multichannel Polarimeter. (With 1 Figure.)

Summary: In the analysis of partially linear polarized light at least three different measurements are necessary to obtain the intensity, the amount of polarization and the plane of vibration. To get all three quantities and an additional fourth value, a special multichannel polarimeter has been designed. The light of a star (or of an extended light source) is measured simultaniously in four parallel telescopes, each of them equipped with color filter, photomultiplier, and an analyzer with its respective plane of vibration in the position angle of 0°, 45°, 90° and 135°. The output of the photomultipliers amplified in a four-channel amplifier is given to four voltage-to-frequency converters, and finally to four counters. After a given integration time, the different counts are scanned, punched on tape and printed by a teletypewriter. In the same way the coordinates, the time of observation, the color filter and other data of importance are recorded. All further steps of reduction are made on a digital computer.

In the analysis of partially linear polarized light at least three different measurements are necessary to obtain the intensity, the amount of polarization and the plane of vibration, or the three Stokes parameters: I, Q, U. This can be done in various ways, for instance

1. in form of three (or more) different samplings in a time sequence, or

2. simultaniously in three (or more) different channels.

Also combinations of both methods are possible. In any case the measurements should be repeated or made in even more channels for higher accuracy. The redundance may serve as an error check.

A differential method to measure the polarization of starlight which can be regarded as a combination of both methods, has been used formerly by the author with good results [1]. In that case, by means of a Wollaston prism, the light was devided into two beams which were polarized in planes of vibration perpendicular to each other. In that way the difference of two channels was measured simultaniously, but settings in different position angles were made successively. The method however, has some disadvantages. The instrumental polarization is not yet compensated, and the result becomes undetermined when the intensity changes during the measuring time.

The method which will be described here, has been developed mainly to measure the polarization of extended light sources, for instance the integrated light of the milky way or the zodiacal light, but it can be used also to measure the polarization of single stars, though other methods may be more efficient in the latter case. The measuring process is automatic to a high degree, the results

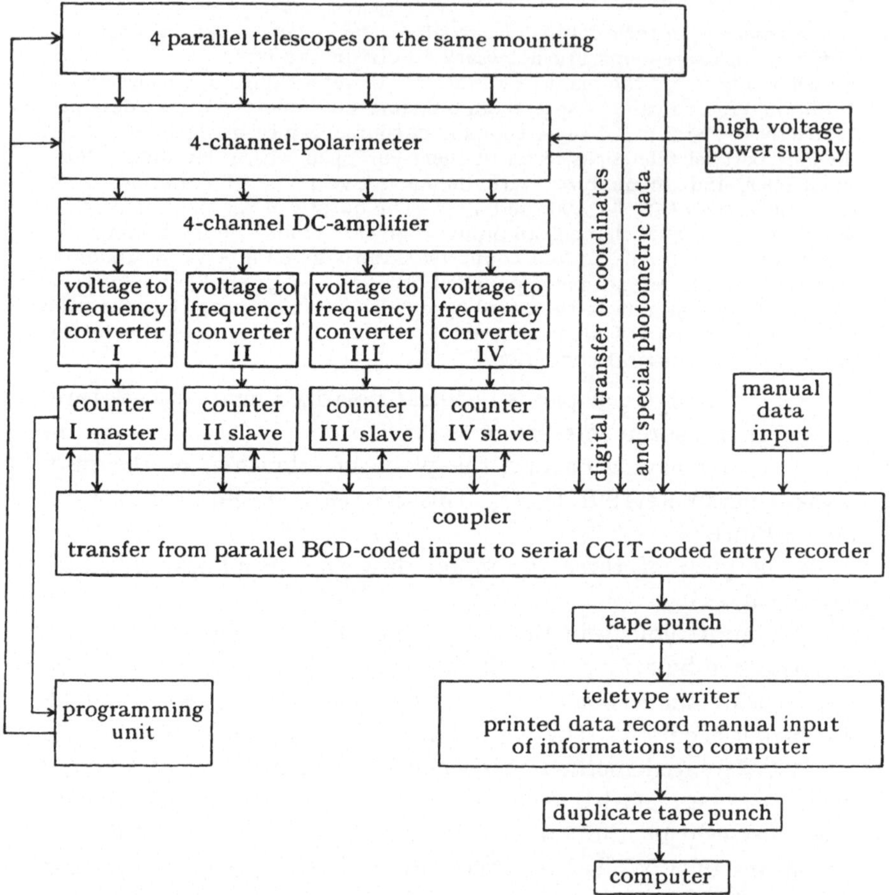

Fig. 1. Block diagram of the designed automatic multichannel polarimeter

are given in digital form. The general arrangement may be demonstrated on the block diagram (Fig. 1).

Four parallel telescopes with primary mirrors of about 25 cm diameter each on a common parallactic fork mounting serve as light receiver. The measured area is determined by diaphragms of 0°.5 and smaller diameter. Four analyzers with their respective planes of vibration in position angles of 0°, 45°, 90° and 135° in the different focal planes of the four telescopes, color filters and photomultipliers are combined to a four-channel polarimeter. The measurements in all channels are made simultaniously. The output of the photomultipliers is finally amplified and given to Hewlett-Packard voltage-to-frequency converters which again are coupled to four electronic pulse counters where the count result is stored in BCD-coded form.

Besides the advantages which are connected with automation and digitizing of the results, the proposed method seems to be superior to former ones for the following reasons:

1. the time of observation is identical in all four channels and therefore

2. seeing conditions, changing transparency of the atmosphere are the same,

3. electric conditions of the photomultipliers (high voltage power supply, supply for the filaments and plate voltages of the amplifiers) are identical,

4. the planes of vibration of the analyzers in all four telescopes are in a fixed position with respect to all other optical parts, instrumental polarization therefore is compensated.

The heart of the automation however, is the dymec-2540 coupler which also receives BCD-coded information about the coordinates of the telescope, special photometric data from the polarimeter (type of color filter used, position of the analyzers, and so on), and manually given data about special events. It transfers all the accepted data from the parallel sources serially to the adjacent tape punch.

The master counter is operated on a quartz time base and can be adjusted to a preset gating time, in this special case mostly an interval of 10 seconds. Any time the counter opens and closes its gate, the gate control signal is also given to the three slave counters so that the actual counting interval is identical to that of the master counter. When all the gates have been closed at the end of the counting interval, an end-count signal is sent to the coupler, as well as to the other input sources, where the buffer storages are brought into "read"-position. The coupler then scans the BCD input signals from the different sources in a given sequence. While the scanner dwells on an input digit, the individual BCD data is translated into international teletype code (CCIT2), and the punch magnets and clutch in the tape punch unit are energized. After each punch cycle the scanner is caused to advance to the next digit for translation and punching. During the time the scanner is in operation, a hold-off signal is sent to each input to provide an indication that recording is in process, thereby ensuring that the counters and input data of the other sources are held until recording is finished. At the end of the complete scanning and recording cycle, a reset signal is given to the counter, the new counting cycle is started, and the scanner returns to its initial position. The whole scanning cycle lasts about 2.5 seconds which is to be compared with the 10 seconds normal gating and counting time. The endcount signal of the master counter is also given to the programming unit where the next step of the special observing program (new coordinates of the telescope, new position of the analyzers and so on) is initiated, the new setting being accomplished during the time that the scanning and punching is in operation. A wide variety of different programs can be controlled on the key-board of the programming unit.

The punched tape is fed to an adjacent teletype-writer where the same data are printed out about a few seconds later. Additional information can be given on the key-board of the writer. A duplicate tape containing the original measurements and further informations, mostly concerning the following computing process, is punched by the teletype writer. This tape finally is used as input to feed a conventional digital computor. All further reductions are programmed in the language of the respective computer.

Literatur

[1] BEHR, A.: Veröffentl. Univ. Stw. Göttingen **114** (1956).

Discussion

Siedentopf: How do you eliminate the difference in sensitivity between the four multipliers?

Behr: At the begin and end of the observations and in reasonable time intervals stars with no, or with well known polarization are measured. To correct for floating sensitivity the measurements are repeated with cyclical interchanging the position angles of the analyzers.

Bahner: What is the advantage of the combination dc-amplifier + voltage-to-frequency converter + counter in comparison to a single pulse counter?

Behr: In our case the photocurrent of the multipliers varies from 10^{-9} to 10^{-7} Amp. In this range the measuring device should have a linear response. A current of 10^{-7} Amp corresponds to about 10^6 pulses/sec. Counters of extremely high resolving power (of the order of 100 Mc/sec) must be used for linear response. In order to measure single pulses a combination of preamplifier + non-overload-amplifier + pulse discriminator + pulse counter is necessary, altogether an expensive equipment, whilst in the proposed case simple industrial counters can be used.

Høg: A remark to the choice between pulse counting and DC-amplification with voltage to frequency conversion: A non-overloading pulse-amplifier with discriminator is the model 212 manufactured by Baird-Atomic (Rohde & Schwarz). It has a delay-line-dipping which makes it very fast. Besides it is cheaper than a voltage-to-frequency converter.

Fricke: Whenever a design for a one-purpose-instrument is being made where one can expect that the observing program will take only a few years, one should clearly state at the very beginning what shall happen to the instrument after the completion of the planned research program for which the instrument was designed. Why don't you plan to use one bigger telescope instead of four smaller ones? The bigger telescope can later be used for a wide variety of other purposes or there might even be a bigger telescope available for your purpose.

Behr: The whole instrumentation as it is described here, can later be used for permanent records of the intensity, color and polarization of the zodiacal light. The electronic part which is most expensive of all, will form a standard equipment for all kind of future multichannel photometry (UBV or other system).

It was planned to use four smaller telescopes instead of one bigger one because it is the only way to observe the necessary quantities simultaniously. Even the splitting of the beam into only two components as in the already cited method (1) is practically not possible in the case of extended light sources. A continuous scanning of the whole sky becomes completely impossible. With only one big telescope instrumental polarization must be determined very precisely or a refractor must be used, in all cases a huge and bulky thing for permanent scanning as to be compared with the proposed relatively small, compact and handy unit. Altogether the main principle of the proposal with its advantages would be abandoned by using only one telescope.

Rohlfs: Why is the first tape-punch included into the unit, and not the telex connected directly to the coupler?

Behr: The coupler together with the first tape punch operates about three times as fast as the teletype writer can do, it therefore serves as a buffer. In addition to this, manually given informations by the type-writer will not disturb the regular scanning and recording process of the coupler.

E. Raimond (Leiden): Digitizing and automatic reduction of 21-cm line observations, made at Dwingeloo, Netherlands. (With 1 Figure.)

Summary: A simple method — now in use since the beginning of 1961 — for recording 21-cm line observations in machine readable form is described. For the standard reduction from receiver output to calibrated observation, the present machine programme improves the ratio between reduction time and observation time by a factor of the order of 10.

Introduction

The 25-metre radio telescope at Dwingeloo, Netherlands, is used for observations of neutral interstellar hydrogen since 1956. The 21-cm line receiver measures intensities as a function of frequency and position (right ascension α and declination δ).

Since the installation in 1958 of an eight-channel receiver, observations are made at a much greater rate than before. The equipment is continuously operated on a routine basis. Per day it produces nearly 5000 independent bits of information. As a result however, the amount of time, necessary for the reduction of the data into a scientifically usable form, has also increased considerably. Hand-reduction of one hour of observation requires at least five hours.

In order to avoid the piling up of unreduced observations a simple digital recording system has been designed, which did not necessitate any major changes to the receiver. We did not aim at the best solution, but at that one which could be realized in the shortest possible time. Therefore, the system described in this paper is only a first step towards full automation of 21-cm reductions. Nevertheless, it served its main purpose by diminishing the ratio between reduction time and observing time by a factor of the order of 10. Moreover, the experiences with this simple and relatively cheap system, which has now been in use since the beginning of 1961, proved to be very important for the planning of further improvements. The twenty-channel parametric receiver, which we hope will be in use towards the end of 1963, is specially designed to have digital output.

Equipment

The eight-channel 21-cm line receiver used for the observations is an improved version of the one decribed by MULLER [1], (see also

AUTOMATIC REDUCTION OF 21-CM LINE OBSERVATIONS

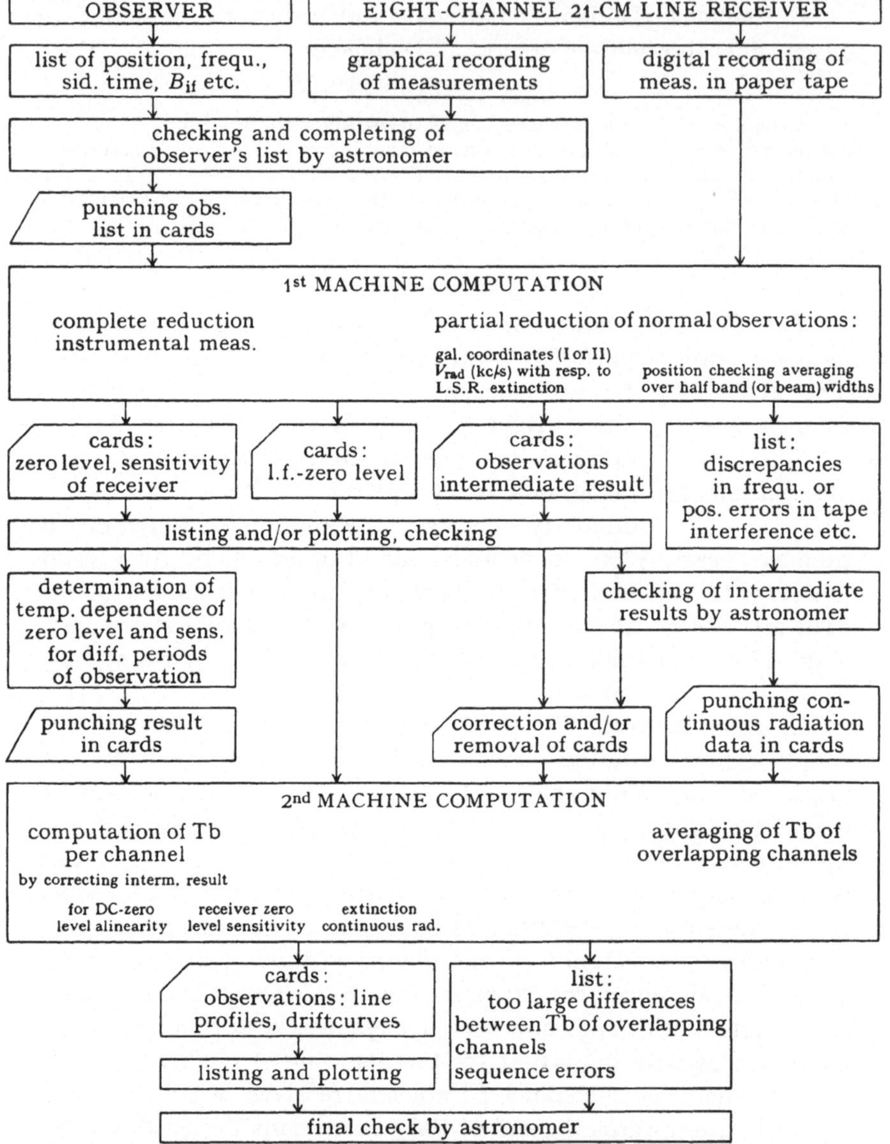

Fig. 1. Block diagram of the automatic reduction of 21-cm line observations

Muller and Westerhout [2]). It measures intensities of the 21-cm line at eight equally spaced frequencies. In order to obtain a line profile all channels are simultaneously moved through the spectrum. The output of the receiver is recorded in analogue form on a Honeywell Brown eight-point recorder and in digital form through a Beckman digital voltmeter and a Creed-25 punch for paper tape. Every six seconds a rotating switch connects the digital voltmeter and the recorder to one of the eight channels. Thus the output of each channel is sampled once every 48 seconds. In order to make the graphical recording readable the zero levels of the eight low-frequency sections are kept unequal.

During a measurement the frequency or the equatorial co-ordinates of the telescope are continuously changed. Whenever the frequency reaches a multiple of 10 kc/s (c.q. α or δ a multiple of 1°) this is marked on the recorder sheet; the paper tape receives a punching in the fifth track, which is not used for punching intensities. For checking purposes multiples of 100 kc/s and 1 Mc/s (c.q. 10° in α or δ) are marked with special punchings.

A serializer, manufactured in the Dwingeloo workshop, edits the tape. It makes sure that all numbers are punched in three rows of one decimal each, and separates consecutive series of eight-channel punchings by a set of three blank rows. At the beginning and end of each measurement it connects the digital voltmeter to an electric thermometer for a measurement of the temperature of the first mixer, and it punches a measurement number and some codes into the tape.

The reduction of the observations requires many more data; at present these are not immediately available in machine-readable form. Therefore, the telescope co-ordinates, the frequencies at the beginning and the end of the measurement, the bandwidth, etc. are listed by the observer to be punched in cards later.

Reduction process

A detailed description of the reduction process, by which the receiver output is converted into brightness temperatures, is in preparation (MULLER, RAIMOND, WESTERHOUT and VAN WOERDEN [3]). The following brief description serves to explain how the process has been adapted for an electronic computer.

1. Reduction of the receiver output.

The output voltage of the receiver must be corrected for:

a) the differences in zero level of the low-frequency sections. These levels are measured at intervals of a few hours;

b) non-linearity of the intermediate-frequency amplification;

c) zero level of the receiver, which is slightly frequency-dependent. For periods of a week or longer it appears possible to relate the height and slope of the zero line to the temperature of the first mixer. Measurements are made once per day;

d) sensitivity of each channel. The relative sensitivities of the eight channels appear to be constant over periods of several weeks while the overall sensitivity is strongly related to the temperature of the first mixer. Usually sensitivity measurements are made once per day;

e) atmospheric extinction;

f) continuous radiation at 21 cm.

2. Reduction of the radio telescope settings.

The horizontally mounted radio telescope is guided by way of a mechanical co-ordinate transformer. The transformer is set in equatorial co-ordinates. Assuming that the errors in the transformation are negligible, the set position still has to be corrected for refraction and for mal-adjustment of the radio-axis of the telescope, both of which are elevation-dependent. In general the results are desired in terms of galactic rather than equatorial co-ordinates.

3. Reduction of the frequency.

The frequency of the tunable local oscillator in the receiver is directly related to the radial velocity of the hydrogen atoms producing the 21-cm radiation. This frequency must, however, be corrected for the effects of the velocity of the earth with respect to the local standard of rest, and of the rotational velocity of the earth.

Electronic reduction

The reduction-process described above is carried out through an electronic computer (Electrologica X1). As shown in the diagram, the reduction consists of two separate machine computations. This separation was necessary because the instrumental data needed for correcting the observations for zero line and sensitivity have to be checked and summarized before they can be used. The measurements from which these data must be determined are interspersed with the normal observations on the tape. Therefore, only the instrumental measurements are completely reduced in the first machine computation. For the other observations the frequency reduction and the position reduction are carried out, and the number of data is decreased by averaging over half bandwidths (c.q. beamwidths in case of drift curves). All results are punched in cards; inconsistencies in the input data found by the machine programme are indicated on a typewriter.

The correction of the observations for variations in the performance of the receiver is carried out in the second machine programme. This programme furnishes brightness temperatures as a function of galactic co-ordinates and radial velocity with respect to the local standard of rest; these are listed, and plotted with a card-reading XY-plotter.

Only at a few stages in this process the astronomer supervising an observing programme has to do a small amount of handwork. He checks and completes the list of data made by the observer by comparing it with the graphical recording of the observations; later he checks the intermediate results after the first machine computation.

Between the first and second machine computation inspection of the instrumental data is necessary. The relations of zero level and sensitivity to the temperature of the first mixer are determined by the method of least squares.

For further scientific interpretation of the observations some machine programmes have been developed. Among those are a programme for computing optical depths and a programme for constructing two-dimensional maps with lines of equal brightness temperatures or equal optical depths. At Groningen, optical depth profiles are analyzed into Gaussian components as described by SCHWARZ and VAN WOERDEN [4] in a paper in this symposium.

Future plans

The parametric twenty-channel receiver to be put into operation in 1963 will increase the speed of observation by another factor of 10. The improvement of the reduction procedure, then necessary, will be mainly achieved by digital recording in paper tape of the telescope co-ordinates, the sidereal time, the date and the frequency, so that the amount of handwork will be reduced considerably. Automation of the settings of telescope and receiver is also under consideration.

References

[1] MULLER, C. A.: Philips Tech. Rev. **17**, 351 (1956). — [2] MULLER, C. A., and G. WESTERHOUT: Bull. Astron. Inst. Neth. **13**, 151, No. 475 (1957). — [3] MULLER, C. A., E. RAIMOND, G. WESTERHOUT and H. VAN WOERDEN: Bull. Astron. Inst. Neth. (in preparation). — [4] SCHWARZ, U., and H. VAN WOERDEN: Sitzungsber. der Heidelberger Akademie, 1962/63, 2. Abh., 107.

K. Rohlfs (Bonn): A 15-Channel Digital Data Recorder for Radio Astronomical Measurements. (With 2 Figures.)

Summary: A description is given of a 15-channel digital data recorder for the 21-cm line receiver of the Stockert radio telescope. The voltages of the 12 receiver channels (10 line channels, one comparison channel for the continuum and one channel for checking) are scanned by a 12-point plotter. The signal at the rotary switch of the plotter is measured by first transforming it into a frequency (voltage-to-frequency-converter Dymec 2210) and then measured by counting the frequency over a fixed time interval (EPUT meter Beckman 7151 H). A digital clock gives the time of the measurement, and two angular position encoders (Hilger and Watts, London) give the elevation and the azimuth of the telescope during the measurement with 4 figures accuracy. All data are punched on an IBM 026 alphanumeric card punch.

The change of the one-channel 21-cm receiver of the Stockert radio telescope to a 10-channel receiver increases the expected output of the telescope in such a way that it is feasible to provide the instrument with a digital output as well. Several considerations, however, restricted the choice of instrument type. The receiver has a rather elaborate control unit which we thought impractical to change. So we planned the digital data recorder as a completely separate unit, which is only an accessory to the receiver and does not impose control upon it. There is, however, some interlocking of receiver function and recorder, so that malfunctioning of the equipment will be detected.

The experience obtained so far with 21-cm measurements indicated the feasibility of providing some additional channels besides the 21-cm line channels proper, viz. provide means for an automatic recording of both telescope position and time of the measurement. So the recorder is able to accept data of three different types: a voltage as given by the integrating network of the receiver, the angular position of a shaft indicating the position of the telescope and the time given by a digital clock. The properties of the control unit of the recorder require the classification of the 15 recorder channels into four classes A, B, C and D.

1. The encoding of the receiver channels A 1—12

The 10 channels of the 21-cm line signal and the fixed comparison channel appear as 11 parallel outputs at the receiver. Our experience has shown that an additional visual display of the measures is advisable, so these voltages are scanned periodically by a 12-point plotter. The signals are scanned on a start pulse given by the receiver control unit (Start A Fig. 1). All integrating circuits are grounded after beeing scanned. The signal appearing at the rotary switch of the plotter is used for the digital recording as well. After some amplification by a FITGO-amplifier (Beckman) the

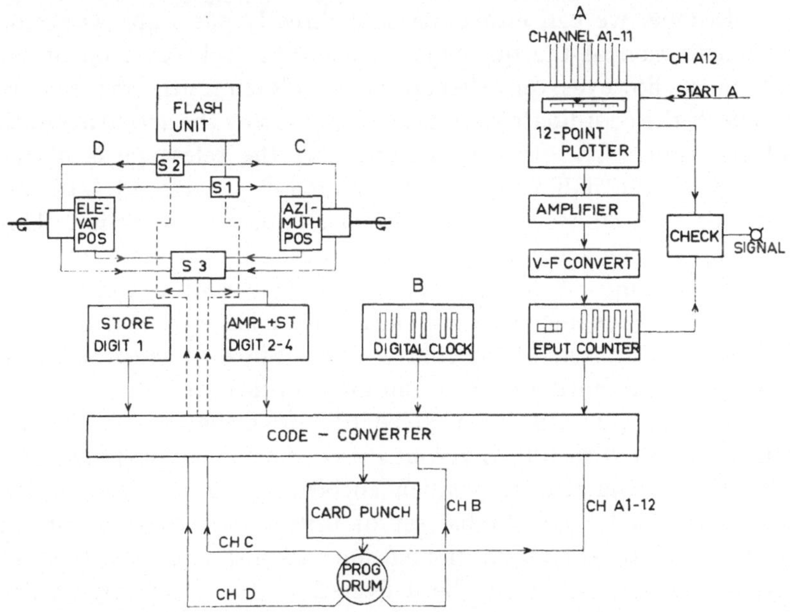

Fig. 1. Simplified block diagram for the digital data recorder

voltage is converted into a frequency by a voltage-to-frequency-converter (Dymec 2210), this frequency then is measured by a counter (EPUT meter 7151 H Beckman).

We preferred this at first sight cumbersome procedure to a simple digital voltmeter since this equipment serves as a perfect integrating circuit as well. In programs, in which extremely long time constants are needed, this type of instrument is simpler and more accurate than any integrating network. The count of the EPUT-meter is transferred via a code-converter to the IBM 026 alphanumeric punch.

2. The measurement of time, channel B

The time is measured by a simple digital clock (Beckman 3203), which is indicating hours, minutes and seconds. No extra time base has been provided for this clock, it is driven from the master oscillator of the receiver. By interposing a solar-time to siderial-time converter the clock directly shows siderial time.

3. *The position of the telescope, channel C and D*

In principle two possibilities exist for measuring the position of the telescope, we can either measure directly the right ascension and declination of the telescope, or azimuth and elevation of the dish. Since, however, the telescope has an altazimunth mounting we can check the coordinate converter by measuring the actual azimuth and elevation. The pointing accuracy of the telescope is of the order of 1′, the half width of the main lobe beeing 15′—16′ for 21-cm, so that an accuracy of 4 decimal figures is called for. This obviously can be achieved in a simple way only by direct angular position encoding. A direct coupling of the coder disks to the axles of the telescope is met with difficulties, they have to be connected to the servo repeaters at the control desk. Since these can provide an angular momentum of 8 gr·cm only, optical encoders have to be used. We selected for each coordinate a combination of an optical encoder with 1000 divisions per revolution coupled mechanically to a mechanical 10-position encoder. Any backlash of the gears is avoided by a double set of brushes for the mechanical encoder. The total angular momentum necessary for driving these units is 5 gr·cm, which can be delivered by the indicating servos. The angular position encoders together with the auxiliary electronic equipment has been delivered by Hilger and Watts, London.

The position encoders of both coordinates are coupled by a relay switch (S_1, S_2, S_3) to the same amplifier-store-code-converter. Thus the coordinates can be measured only one after the other, but since the measurement and printing of one coordinate takes only approx. 300 msec, this should be of no importance. The measurement of the coordinate itself takes only 2 msec since the coder disks are flashed. Thus the measurement can be made even if the telescope is moving with highest speed.

4. *Control of the digital recorder*

The digital recorder described here is programmed partly by the second program drum of the IBM 026 punch. The other part of the program is fixed. By punching holes in the program card one can select, which of the four main channels A, B, C or D is to be activated, and which of the decades 1 to 6 is to be punched. It is possible to punch the counter values in a perfect arbitrary order and to omit certain decades.

If, however, the 12-point plotter is started by "start A", it begins scanning all 12 channels A 1 to A 12, the program drum then only can select, which of these channels is to be printed. Fig. 2 gives a simplified pulse plan for the control of channel A. Start activates the plotter and it begins scanning the 12 channels. From the plotter we derive a signal, which marks the times, in which the wiper of the rotary switch is touching one of the contacts. From this signal we obtain "contact begin" and "contact end" pulses. The "contact begin" pulse is the start pulse for the counter, the end of the counting

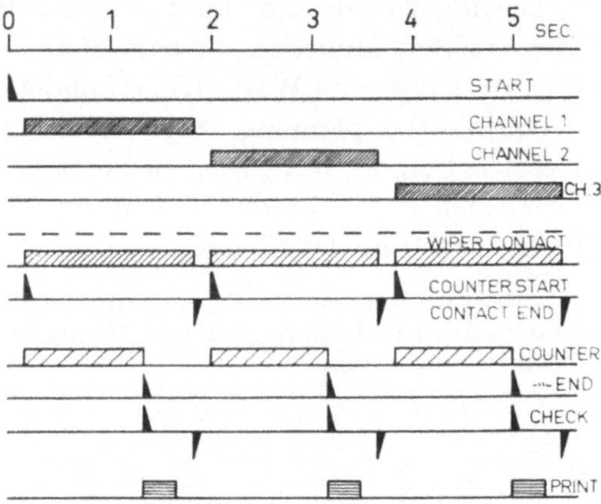

Fig. 2. Pulse plan of the control for scanning channels A 1—12

period is governed by the time base of the counter. Since the counting interval is set manually on the counter itself, it is possible to make program errors, if one chooses the counting interval longer than the time, the wiper has contact. This is checked by a check circuit, and a visual and audible signal is given, as soon as the counting interval is too long. "Counter end" activates the printer and the count is printed. Again it is possible to select, which of the decades is to be printed.

Channels B, C, and D are controlled by the card punch and do not need an external start pulse. If on the program card channel C is called, first the switches S_1, S_2 and S_3 are set, then the flash unit is fired and thereby the position reading in one coordinate is taken. The measured value then is automatically transferred to the card. In a similar way the clock reading is taken.

It is possible to delay the position and time measurements until an external signal has arrived, but for 21-cm measurements this is not intended.

The digital data recorder has been planned in such a way that it will be possible to add more channels later on and even to transfer some parts of the control of the receiver to it. This will have the advantage, that these functions then can be card-programmed. Most probably the first step into this direction will be, to take the frequency setting of the receiver from the cards. We will wait, however, for experiences with the data recorder before taking this step.

The main contractor for the digital data recorder is Beckman Instr. GMBH, München. A part of the equipment has been delivered by Hilger and Watts LTD, London and by Hewlett-Packard.

During the planning stage of the recorder discussions with several colleagues have been of great help, my special thanks are due to Dr. B. H. Grahl of Universitäts Sternwarte Bonn and Mr. E. Høg from Hamburger Sternwarte.

Note added in proof: In the meantime the equipment has been delivered and is installed at the radio-telescope. A description of the complete receiving unit is in preparation.

H. G. Girnstein (Bonn): Erfahrungen mit der Approximation von beobachteten 21 cm-Profilen durch Gaußkurven.

Summary: The Gaussian analysis of 21-cm profiles by the method of least squares is met with difficulties caused by the small radius of convergence. It is shown how this obstacle can be overcome by splitting the solution of the normal equations into three steps instead of the usual single one. Some first results of this method in the Monoceros region are discussed. By the application of the Gaussian analysis it was possible to obtain from the 21-cm line measurements some indications on the existence of an expanding shall of neutral hydrogen in the association I Mon.

Zweck und physikalische Voraussetzungen der Methode

Unter „Gaußanalyse" verstehen wir die Zerlegung einer Kurve in eine Summe von Gaußschen Glockenkurven. Physikalisch sinnvolle Aussagen sind aus dieser Art der Analyse nur dann zu entnehmen, wenn man Gründe zur Annahme hat, daß die Meßkurve auch wirklich aus einer Überlagerung einzelner Gaußkurven entstanden ist.

Wir legen der Interpretation der Meßdaten das folgende Modell zugrunde: Die 21 cm-Strahlung entsteht in einzelnen HI-Regionen. Diese HI-Regionen können entweder aus einem einzelnen Turbulenzelement bestehen, oder aber größere Aggregate sein, die aus vielen solchen Einzelelementen zusammengesetzt sind. Wenn die erste Möglichkeit verwirklicht ist — wie es z. B. in hohen galaktischen Breiten der Fall sein kann — wird die Linie nur durch die thermische Bewegung der Atome verbreitert, ist also eine Gaußkurve. Im zweiten Fall braucht dies nicht unbedingt der Fall zu sein, da die Geschwindigkeitsverteilung der einzelnen Turbulenzelemente von einer Gaußverteilung abweichen könnte. Dies wird z. B. von BLAAUW [1] behauptet. Nach Untersuchungen der interstellaren Na- und Ca II-Linien von SCHLÜTER, SCHMIDT und STUMPFF [2] ergibt sich aber eine Gaußkurve für die Geschwindigkeitsverteilung, wenn man die circumstellaren Linien ausschließt. Daher glauben wir annehmen zu können, daß auch das interstellare Wasserstoffgas eine Gaußverteilung in den Turbulenzgeschwindigkeiten zeigt.

Sind die Unterschiede der Relativgeschwindigkeiten der HI-Regionen so groß, daß sich ihre Strahlungskomponenten nicht allzusehr überlagern, so kann man das gemessene Profil

$$y = P(v) \tag{1}$$

mit den diskreten Werten P_i und v_i durch eine Funktion

$$y = \sum_{l=1}^{n} H_l \exp - \left(\frac{v - a_l}{b_l}\right)^2 \qquad (2)$$

darstellen. Diese Darstellung kann zu inkorrekten Ergebnissen führen, wenn die mittleren Radialgeschwindigkeiten der HI-Regionen zu wenig verschieden sind, da dann die Form der Emissionsprofile wegen der Absorption in vorgelagerten Wolken von Gaußkurven abweichen kann. Dies ist besonders in Richtung auf Zentrum und Antizentrum der Galaxis der Fall. In den meisten anderen Fällen ist eine solche Schwierigkeit nicht zu befürchten. Sollten die optischen Dichten so groß werden, daß sie berücksichtigt werden müssen, kann dies in der hier beschriebenen Methode mit nur wenig größerem Aufwand geschehen.

Die Berechnung der Parameter H, a und b

Da es sich bei der Gl. (2) um eine Funktion mit nichtlinearer Regression handelt, muß zur Bestimmung der Parameter ein Näherungsverfahren angewandt werden. Wir benutzten hierzu die Methode der kleinsten Quadrate.

Die Normalgleichungen der Koeffizienten von Gl. (2) stellen wir auf nach der bekannten Methode (siehe z. B. Rietz-Baur [3]) durch Multiplikation der beiden Matrizen (3).

$$\begin{vmatrix} \frac{\partial y_1}{\partial H_1} & \frac{\partial y_2}{\partial H_1} & \cdots & \frac{\partial y_k}{\partial H_1} \\ \frac{\partial y_1}{\partial a_1} & \frac{\partial y_2}{\partial a_1} & \cdots & \frac{\partial y_k}{\partial a_1} \\ \frac{\partial y_1}{\partial b_1} & \frac{\partial y_2}{\partial b_1} & \cdots & \frac{\partial y_k}{\partial b_1} \\ \vdots & \vdots & & \vdots \\ \frac{\partial y_1}{\partial b_n} & \frac{\partial y_2}{\partial b_n} & \cdots & \frac{\partial y_k}{\partial b_n} \end{vmatrix} \times \begin{vmatrix} \frac{\partial y_1}{\partial H_1} & \frac{\partial y_1}{\partial a_1} & \frac{\partial y_1}{\partial b_1} & \cdots & \frac{\partial y_1}{\partial b_n} & \alpha_1 \\ \vdots & \vdots & \vdots & & \vdots & \vdots \\ \vdots & \vdots & \vdots & & \vdots & \vdots \\ \frac{\partial y_k}{\partial H_1} & \frac{\partial y_k}{\partial a_1} & \frac{\partial y_k}{\partial b_1} & \cdots & \frac{\partial y_k}{\partial b_n} & \alpha_k \end{vmatrix} = \begin{Bmatrix} a_{il}\,(H_l, a_l\,b_l) \\ l = 1, 2, \ldots n \end{Bmatrix} \cdot (3)$$

Hierbei sind:

$$\left. \begin{aligned} \frac{\partial y_i}{\partial H_l} &= e^{-\left(\frac{v_i - a_l}{b_l}\right)} \\ \frac{\partial y_i}{\partial a_l} &= 2 \cdot \left(\frac{v_i - a_l}{b_l}\right) \cdot \frac{1}{b_l} \cdot H_l \cdot e^{-\left(\frac{v_i - a_l}{b_l}\right)^2} \\ \frac{\partial y_i}{\partial b_l} &= 2 \cdot \left(\frac{v_i - a_l}{b_l}\right)^2 \cdot \frac{1}{b_l} \cdot H_l \cdot e^{-\left(\frac{v_i - a_l}{b_l}\right)^2} \\ \alpha_i &= P_i - y_i \end{aligned} \right\} \qquad (4)$$

Die rechte Seite von (3) ist das Koeffizientenschema für das Gleichungssystem:

$$\left.\begin{array}{l}
a_{(1)(1)}\,\Delta H_1 + \; a_{(1)(2)}\,\Delta a_1 + a_{(1)(3)}\,\Delta b_1 + \cdots a_{(1)(3n)}\,\Delta b_n + \quad a_{10} = 0 \\[4pt]
a_{(2)(1)}\,\Delta H_1 + \; a_{(2)(2)}\,\Delta a_1 + a_{(2)(3)}\,\Delta b_1 + \cdots a_{(2)(3n)}\,\Delta b_n + \quad a_{20} = 0 \\[4pt]
\cdots \cdots \cdots \cdots \cdots \cdots \cdots \cdots \cdots \cdots \cdots \cdots \cdots \cdots \cdots \cdots \cdots \cdots \\[4pt]
a_{(3n)(1)}\,\Delta H_1 + a_{(3n)(2)}\,\Delta a_1 + a_{(3n)(3)}\,\Delta b_1 \cdots a_{(3n)(3n)}\,\Delta b_n + a_{(3n)(0)} = 0.
\end{array}\right\} \quad (5)$$

Die Unbekannten im Gleichungssystem sind die Verbesserungen der Parameter. Obwohl wir formal ein lineares Gleichungssystem erhalten, ist das Problem nichtlinear, da die Koeffizienten des Gleichungssystems nicht konstant sind, sondern von den Parametern selbst abhängen.

Es ist jetzt naheliegend, die Auflösung des Gleichungssystems (3—5) so zu versuchen, daß man mit einer Annahme für H, a und b in (3) und (4) eingeht und dann nach (5) die Verbesserungen der Parameter berechnet. Das Verfahren wird wiederholt und sollte zur Konvergenz führen. Es läßt sich zeigen, daß diese Methode konvergiert, wenn die Ausgangsnäherung innerhalb eines gewissen Konvergenzbereiches liegt.

Bei der numerischen Durchrechnung des Verfahrens war Konvergenz nur selten zu erreichen, da der Konvergenzbereich sehr klein ist, und die vorzugebenden Parameter H, a und b nur schlecht geschätzt werden können. Es zeigte sich, daß nur für ungefähr 5 % der Ansätze die Rechnung konvergierte. Um langwieriges Probieren zu vermeiden, haben wir das Verfahren abgeändert.

In der Ableitung $\partial y/\partial H$ ist H selbst nicht enthalten, so daß für die Bestimmung von H bei vorgegebenen a und b das Verfahren linear ist und H in einem Rechengang bestimmt werden kann. Mit diesem schon relativ gutem Wert von H wird dann schrittweise erst a und dann b berechnet. Weichen die alten und die neuen Werte voneinander ab, wird das Verfahren wiederholt. Die Auflösung geht also nach folgendem Schema vor sich:

$$H_{(a,b)} \Rightarrow a_{(H,b)} \Rightarrow b_{(H,a)} \Rightarrow H_{(a,b)} \Rightarrow \text{etc.}$$

Dabei werden die in Klammern gesetzten Parameter in dem betreffenden Rechengang nicht variiert.

Damit das Verfahren nicht bei den kritischen ersten Schritten divergiert, wurden für die Parameterverbesserungen Δa und Δb Höchstwerte von 2 km/sec zugelassen. Mit dieser Methode wurden

30—40 Iterationen benötigt, d. h. die Rechnung wurde abgebrochen, wenn sich der Wert der normalisierten Fehlerquadratsumme

$$\varepsilon = \sum_{i=1}^{k} \frac{d_i^2}{k} \tag{6}$$

(d_i = berechneter Funktionswert — gemessenem Funktionswert, k = Zahl der Meßwerte)

bei zwei aufeinanderfolgenden Näherungszyklen um weniger als $(0{,}05° \text{ K})^2$ unterscheidet.

Die lange Rechenzeit konnte dadurch erheblich verkürzt werden, daß die Parameterverbesserungen, wenn sie eine bestimmte Grenze unterschritten, mit einem geeigneten Faktor multipliziert wurden. Hierbei mußte darauf geachtet werden, daß kein nennenswertes Oszilieren um das gewünschte Ergebnis eintrat.

Die Gaußanalyse nach beiden Verfahren wurde für den in Bonn verfügbaren Rechner „Standard Elektrik ER 56" programmiert. Nach dem ersten Verfahren, das nach einigen Proberechnungen aufgegeben wurde, betrug die Rechenzeit für ein Profil mit 80 Meßpunkten bei 6 Komponenten: 7 Iterationen a 5 min gleich 35 min.

Das zweite, abgewandelte Verfahren benötigte für den gleichen Fall 12 Iterationen a 2 min gleich 24 min. Es hat also neben dem Vorteil des größeren Konvergenzradius auch noch den Vorzug einer kürzeren Rechenzeit. Dies ist hauptsächlich bedingt durch die Verkleinerung des Gleichungssystems gegenüber dem ersten Verfahren. Auch ist der Bedarf an Speicherplatz für diese Methode wesentlich geringer als beim ersten Verfahren.

Die Analyse gemessener Profile

Bei der Anwendung des oben geschilderten Verfahrens auf gemessene Profile tritt eine weitere Schwierigkeit auf. Die Ordinaten y_i sind mit einem Meßfehler behaftet, der nicht vernachlässigbar ist. Dies bedingt dann, daß es keine eindeutige Darstellung (2) der Funktion (1) mehr gibt. Wir mußten daher Kriterien suchen, um rein formale Lösungen von (2) auszuscheiden.

Folgende drei Kriterien wurden bei der Analyse berücksichtigt:

1. Der Mittelwert der Abweichung der Darstellung (2) von der gemessenen Funktion (1)

$$f = \sum_{i=1}^{k} \frac{d_i}{k}$$

soll die Größenordnung des Meßfehlers haben (etwa 1° K).

2. Sind mehrere Lösungen möglich, die Bedingung 1 erfüllen, dann ist diejenige zu wählen, welche die wenigsten Komponenten besitzt.

3. Zeigen sich bei der Zerlegung eines Profils Ähnlichkeiten der Komponenten mit denen benachbarter Punkte am Himmel, so wird dies als Bestätigung der Resultate aufgefaßt.

Vor der praktischen Anwendung der Gaußanalyse wurde die Methode an synthetischen Profilen mit künstlichem Fehler, der dem der Apparatur entspricht, geprüft. Dabei ergab sich als mittlerer Fehler des Rechenverfahrens für

$$H: \pm 1{,}8° \text{ K}; \; a: \pm 1{,}0 \text{ km/sec}; \; b: \pm 0{,}7 \text{ km/sec.}$$

Nach diesem Verfahren wurde ein Gebiet im Monoceros von 96 Quadratgrad Ausdehnung (86 Meßpunkte) untersucht. Da durch die Gaußanalyse der Informationsgehalt eines 21 cm-Spektrums in einigen wenigen Parametern ausgedrückt wird, ist es relativ einfach, die Verteilung des Wasserstoffgases in einem Gebiet zu analysieren. Für das untersuchte Areal ergaben sich relativ zwanglos drei Spiralarme. Wie sich schon bei einer ähnlichen Untersuchung von Grahl [4] ergeben hatte, zeigten diese Arme teilweise Aufspaltungen, deren Ursache noch nicht geklärt ist. Der Verlauf der Kammlinien bei unserer Untersuchung ist aber wesentlich glatter als bei der Grahlschen Untersuchung. Dies dürfte an der genaueren Analyse der hier beschriebenen Untersuchung liegen.

Von dieser Aufspaltung der Spiralarme gut zu unterscheiden sind andere lokale Strukturen. Besonders in den Breiten $b^{\mathrm{I}} = +1$ und $b^{\mathrm{I}} = +2$ findet man eine Aufspaltung des ersten Spiralarms zwischen $179° > l^{\mathrm{I}} > 173°$. Auch der dritte Arm zeigt an dieser Stelle ein ganz ähnliches Verhalten. Da in diesem Gebiet die Assoziation I Mon liegt, war es naheliegend, diese Aufspaltung damit in Verbindung zu bringen. Wir machten daher ein einfaches Modell dieser Assoziation durch eine expandierende Kugelschale aus neutralem Wasserstoff und versuchten die Abweichungen in dem beschriebenen Gebiet dadurch darzustellen. Hierbei ergab sich eine recht gute Übereinstimmung mit den optischen Daten der Assoziation (Schmidt-Kaler u. van Schewick [5]).

Da ein wirklich stichhaltiger Nachweis atomaren Wasserstoffs in expandierenden Assoziationen sehr wichtig ist, werden die Messungen in der Gegend der Assoziation noch weiter fortgesetzt. Es soll versucht werden, mit Hilfe des Assoziationsmodells auch die

beobachteten Intensitäten der einzelnen Komponenten zu erklären.
Dieser Teil der Untersuchung ist jedoch noch nicht abgeschlossen.

Meßgröße	Assoziationsmodell	
	radio-astronomisch	optisch-astronomisch
Abstand des Expansionszentrums v. d. Sonne	1,6 kpc	1,6 kpc
Expansionsgeschwindigkeit	22,6 km/sec	34 km/sec
Durchmesser	178 pc	140 pc
lineares Expansionsalter	$3,8 \times 10^6$ a	$2,8 \times 10^6$ a
Koordinaten des Zentrums l^{I}	$175{,}^{\circ}8$	$175{,}^{\circ}9$
b^{I}	$+1{,}^{\circ}1$	$+0{,}^{\circ}9$

Literatur

[1] Blaauw, A.: Bull. Astron. Inst. Neth. 11, 405, No. 433 (1952). — [2] Schlüter, A., H. Schmidt u. P. Stumpff: Z. Astrophys. 33, 194 (1953).— [3] Ritz-Baur: Handbuch der mathematischen Statistik. Berlin u. Leipzig 1930. — [4] Grahl, B. H.: Forschungsber. des Landes NRW, Nr. 423 (1960). — [5] Schmidt-Kaler, T., u. H. van Schewick: In Vorbereitung.

Discussion

Wilhelmsson: I would like to remark to the talk by Mr. Girnstein. In fact, the arms in the anticenter region can be resolved also without using a gaussian analysis as demonstrated in recent work by Dr. B. Høflund, at Chalmers University of Technology, Gothenburg.

Lindblad: If the Kootwijk measurements are plotted in the form radial velocity versus galactic longitude I think that you in the same region also can recognize three heavy spiral arms.

U. Schwarz and **H. van Woerden** (Groningen): Electronic analysis of 21 cm line profiles into Gaussian components. (With 2 Figures.)

Summary: Decomposition of spectral line profiles into a few Gaussian components is carried out on the ZEBRA electronic computer. The method is described with particular emphasis on the criteria for choosing the best out of several possible solutions. The program is suitable for application in several other fields of research.

Introduction

At many points of the sky, measurements of the 21-cm spectral line emitted by neutral interstellar hydrogen give profiles evidently containing several components. Since 1959, members of the Kapteyn Laboratory staff have studied the structure of such profiles by analyzing them into Gaussian components with the aid of an electronic computer. KAPER [1] has described an early version of the computer program. A discussion of a later version is now being prepared (KAPER, SMITS, SCHWARZ, TAKAKUBO and VAN WOERDEN [2]); a flow diagram can already be made available. In the present paper we only briefly discuss the method of analysis, which is closely similar to the one described by GIRNSTEIN, but we pay special attention to the criteria used in comparing several possible solutions of a problem.

Method and Computer program

The measurements obtained at Dwingeloo Radio Observatory, after some corrections — cf. the paper by RAIMOND [3] in this Symposium —, give brightness temperature T_b as a function of frequency v. Optical depths τ have the advantage over T_b of being additive; in converting T_b into τ, we assume the kinetic temperature of neutral hydrogen to be $125°$ K. We may then represent the profile $\tau(v)$ by the sum of a few components, tentatively to be identified with hydrogen clouds, each of which is assumed to have a Gaussian frequency (or velocity) distribution:

$$\tau_k(v) = \sum_{i=1}^{n} \tau_{0i} \exp\left[-\frac{(v_k - v_{0i})^2}{2\,\sigma_i^2}\right] + \Delta y_k.$$
$$k = 1, 2, \ldots, m.$$

In this equation, m is the number of observed points in the profile; τ_{0i}, ν_{0i} and σ_i are the parameters (central optical depth, central frequency and dispersion) of the Gaussian components, and n is the number of components.

Since the measured values τ_k are affected by receiver noise and other errors, the representation of the profiles cannot be exact, but the best solution may be sought by application of the least-squares method:

$$S = \sum_{k=1}^{m} (\Delta y_k)^2 \text{ should be a minimum,}$$

where Δy_k is the difference of the observed value τ_k and the corresponding ordinate of the computed curve. It is hardly possible to find the parameters x_l ($l = 1, 2, \ldots 3n$; $x_l = \tau_{0i}$, ν_{0i} or σ_i) directly from the equations $\partial S / \partial x_l = 0$; therefore, we start with an estimated set of parameters, expand $\partial S / \partial x_l$ into Taylor series and keep only the first-order terms in Δx_l; the linearized normal equations then yield corrections Δx_l to the estimated parameters. As a consequence of the neglect of the higher-order terms, the corrections Δx_l do not immediately give the best solution; iteration of the correction procedure is necessary in order to approach the minimum value of S. The corrections Δx_l may even lead to an increase in S. Such a "divergence", caused by the non-linearity of the problem, may be overcome by applying only a fraction A of the calculated corrections to the parameters. Takakubo has proved that an interpolation factor $0 < A < 1$ giving a decrease in S does always exist. Interpolation is also useful if one of the parameters τ_{0i} or σ_i becomes negative.

The program for the Stantec-ZEBRA computer was developed by H. G. Kaper and D. W. Smits of the Mathematical Institute of Groningen University; we gratefully acknowledge their work. This program provides for a choice of automatic or manual interpolation; in the former case, any initial estimate will lead to a convergent iteration, although the process may be slow. Iteration continues until two requirements are fulfilled: between two consecutive approximations, the relative decrease of S should be smaller than a given constant ε, and the corrections to all parameters should be smaller than η times their mean errors (as calculated from the normal equations), η being another given constant. Usually we take $\varepsilon = 0.001$ and $\eta = 0.5$, the ε requirement being the stricter.

In order to speed up the calculation of the elements of the determinant of the normal equations, the summation of products of the first derivatives of the Gaussian components is replaced by an integration. After each iteration step, the new approximated set of parameters and corresponding S are printed. At the end of the iteration process, the observed and computed values of τ_k and their difference are punched and listed; besides, the machine supplies the final values of the parameters, the quantities N_{Hi} (number of neutral hydrogen atoms per cm^2 in a component, proportional to $\tau_{0i} \times \sigma_i$), and the mean errors of these quantities. The computer requires about 1.25 m n seconds of time per iteration step; it accepts a maximum of 6 components and 150 measured profile points.

Criteria for judging solutions

Because of the random fluctuations in τ_k, different solutions of $\partial S/\partial x_l = 0$ with the same number of components may exist; and because of the nonlinearity of the problem, different initial estimates do not always lead to the same solution. In Fig. 1, we show two examples of profiles for which two different solutions of comparable quality are available. Since, moreover, the number of components is not known a priori, we usually calculate a few solutions with different numbers of components.

In order to decide which solution among several calculated is the best — in other words, the most probable — representation of the profile, we use two main criteria; one is the distribution of residuals Δy_k, the other is the size of the relative mean errors in the parameters. Let us consider these criteria in some detail.

The primary criterion concerns the residuals. Their distribution should be random, their size in reasonable agreement with noise and other observational errors. In a good solution, no systematic deviations should be present; if they are, they may indicate an extra component. To judge the size of the residuals, we compute Q, the proportion between S and an expected value S_{th}. The estimate of S_{th} is derived from the noise errors in τ_k; from many profile measurements, we have taken the fluctuations in the far ends, where radiation is negligible, as the basis of our noise statistics. The value of Q should not be far above unity; neither should it be appreciably lower. A further useful quantity is the fraction g of large residuals (here defined as residuals exceeding twice the mean noise error). In summary, the distribution of $\varDelta y_k$, or the size of Q or g, may indicate the necessity of including an extra component — or suggest that one should be dropped or two combined, in case too good a fit has been obtained. The latter indication, however, usually is not clearly recognizable; in such cases, the second criterion is particularly important.

The secondary criterion relates to the mean errors of the parameters. We require v_i, the sum of the relative mean errors in τ_{0i} and σ_i, to be smaller than 0.50. If $\tau_{0i} \times \sigma_i$ is small, larger values of v_i may occur, for example in a component which only serves to represent a peak in the noise fluctuations; we note that the instrumental bandwidth imposes a lower limit on the dispersion σ_i of acceptable Gaussian components. Values of v_i may also be large, if a strong correlation between parameters of two or more components makes some of the parameters highly uncertain; an example

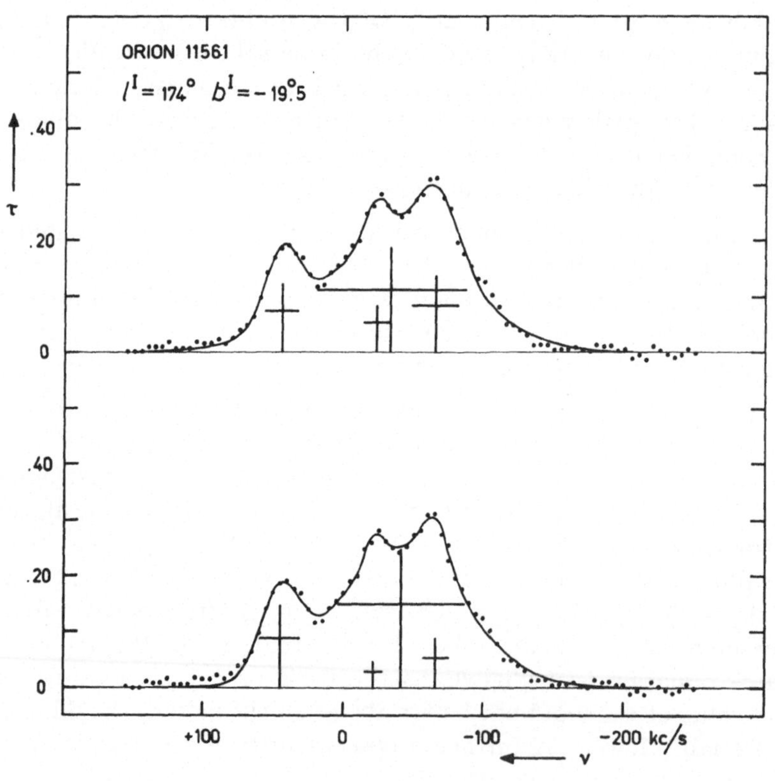

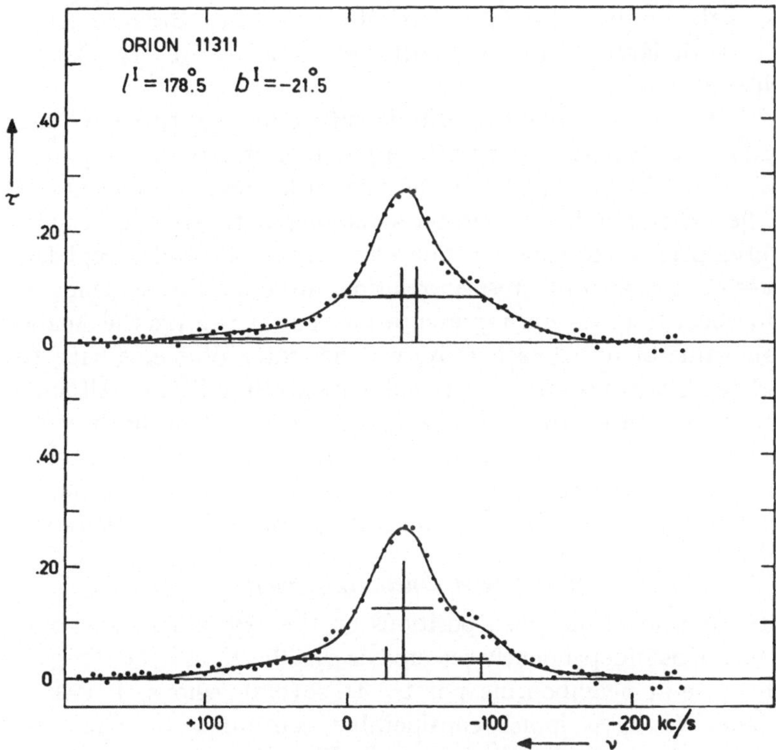

Fig. 1. Examples of ambiguity (see also page 110)

The analysis of profile numbers 11 561 as well as 11 311 led to two solutions each of almost equal quality. Measured values of τ_k are shown as dots, the sum of computed components as a continuous curve. Each component is represented by a cross of height τ_{0i}, erected at the abscissa v_{0i}, and with $2\sigma_i$ as cross-beam.

The quantities Q, g and $\overline{v_i}$ defined in the text have the following values:

	Q	g	$\overline{v_i}$
11 561 upper	1.92	0.16	0.15
11 561 lower	1.86	0.15	0.21
11 311 upper	1.86	0.14	0.24
11 311 lower	1.88	0.17	0.19

For no. 11 561, the lower solution should be rejected because of systematic residuals around $+100$ kc/s. In no. 11 311, one might choose the solution with the smaller value of Q.

is a weak component on the slope of a wide one. Because components with large v_i are of questionable reality, we try to avoid such solutions.

In most cases, these two criteria suffice for selecting the "best" solution. If two solutions with different numbers of components fulfil both criteria, we choose the one with fewer components. If, on the other hand, no solution satisfying both criteria is found, we give priority to consideration of the residuals and accept large values of v_i caused by cross-correlation, although the tendency will be to select from several representations one that gives the smallest mean value of v_i. Occasionally, we encounter problems with two nearly equivalent solutions; examples are given in Fig. 1. Although, even in such cases, rules may be laid down for choosing the better representation, the nonuniqueness of the solution to the problem of Gaussian analysis then receives a strong emphasis. From our experience, these ambiguous problems appear to be relatively rare.

Results and concluding remarks

If the grid of observed positions on the sky is narrow enough, the solutions independently selected as the "best" may be compared with those at neighbouring points. In several regions, Takakubo [4] and we have found considerable continuity of component structure between neighbouring profiles, indicating large-scale structure of the hydrogen clouds. As an example, we present in Fig. 2 a small selection of results obtained in the Orion region (van Woerden [5]). Not everywhere does continuity prevail. We have found that such failure is often caused by the non-uniqueness of the solutions. Re-analysis starting from estimates based on results in the neighbourhood may prove successful. Moreover, in building up a map of cloud structure in a particular region, solutions may be selected which originally were second choice but happen to give a more coherent picture.

Clearly, the method of analysis lends itself to applications to a wide variety of problems. The ZEBRA program developed by Kaper [1] has already been adapted to analysis of the solar helium spectrum at Utrecht Observatory. But multi-Gaussian distributions abound in many fields of research.

Finally, our discussion of the criteria, used in choosing between the several solutions possible in a non-least-squares problem, should serve another purpose: whereas a more rapid production

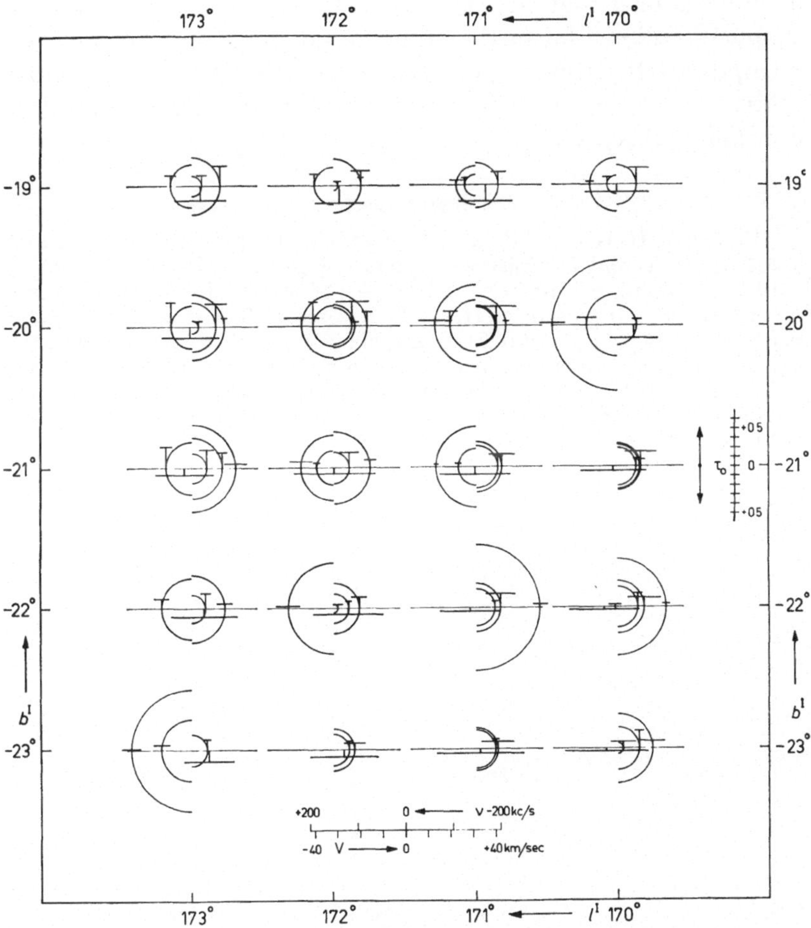

Fig. 2. A preliminary summary of profile components in part of the Orion region
(van Woerden [5])

At each of 20 points on a 1°-grid in galactic coordinates (l^I, b^I), component structure of the 21-cm profile is shown. The central frequency ν_0 of a component is represented by the radius of a half-circle; the central intensity τ_0 by the length of a vertical line tangent to the half-circle; the dispersion σ by half the length of the corresponding cross-beam. The scales of ν_0 and σ are given in the bottom of the figure, in frequency and in velocity measure; the scale of τ_0 is shown at right.

Components with $\sigma > 50$ kc/s are drawn downwards, and their frequency-circles omitted; components with both $\sigma > 100$ kc/s and $\tau_0 < 0.025$ are omitted completely.

of observational material, and availability of a faster electronic computer, will call for further automation of the method, we want to emphasize the great importance of a highly critical judgment of solutions obtained; such judgment is essential to the acquisition of meaningful results.

References

[1] Kaper, H. G.: A method for fitting the best superposition of Gaussian distributions through measured points. Report TW-1, Mathematical Institute, Groningen University (1959). — [2] Kaper, H. G., D. W. Smits, U. Schwarz, K. Takakubo and H. van Woerden: Bull. Astron. Inst. Neth. (in preparation). — [3] Raimond, E.: Symposium on Automation and Digitalization in Astronomy. Sitzgsber. Akad. Heidelberg (1962). — [4] Takakubo, K.: Bull. Astron. Inst. Neth. (in preparation) (1963). — [5] Woerden, H. van: De Neutrale Waterstof in Orion. Thesis, Groningen University, 1962. Bull. Astron. Inst. Neth. (in preparation).

H. Wilhelmsson (Göteborg): The problem of de-smoothing of
21 cm observations. (With 4 Figures.)

Summary: The problem of de-smoothing of data without affecting the
observational errors is treated under the assumption of a gaussian smoothing
function. For special cases a technique of de-smoothing by solving the
integral equation governing the smoothing exactly for simple local approximations to the observed curve is suggested.

1. Introduction

In several fields of physics and astronomy, for example in
spectral optics, stellar statistics, or radio astronomy, the results
attainable by direct observations are often data which are smoothed
with respect to an independent variable.

The integral equation governing the smoothing may be written

$$\Phi(\omega) = \int_{-\infty}^{\infty} K(\omega-\omega')\,\Psi(\omega')\,d\omega', \tag{1}$$

where $\Phi(\omega)$ is the observed function and $\Psi(\omega')$ is the de-smoothed
function to be determined, assuming a certain kernel $K(\omega-\omega')$,
which describes the smoothing effect.

As early as 1910 K. SCHWARZSCHILD in a paper [1] pointed out
that the integral equation (1) may result in negative values of the
de-smoothed function when applied to stellar statistical observations
of star densities.

A method suggested by A. S. EDDINGTON [2], [3] in 1913 has
been applied extensively to radio astronomy as well as to stellar
statistics in order to obtain approximate solutions of the integral
equation (1) in the case of a gaussian kernel, i.e. where

$$K(\omega-\omega') = \frac{1}{\sqrt{2\pi}}\,\frac{1}{\sigma}\,e^{-(\omega-\omega')^2/2\sigma^2}, \tag{2}$$

and σ is the dispersion.

The result obtained by the Eddington method [2] is the
following:

$$\left.\begin{aligned}
\Psi(\omega) = e^{-\frac{\sigma^2}{2}\frac{d^2}{d\omega^2}}\,\Phi(\omega) = \Phi(\omega) - \frac{\sigma^2}{2}\,\frac{d^2\,\Phi(\omega)}{d\omega^2} + \frac{1}{2!}\left(\frac{\sigma^2}{2}\right)^2\frac{d^4\,\Phi(\omega)}{d\omega^4} - \\
- \frac{1}{3!}\left(\frac{\sigma^2}{2}\right)^3\frac{d^6\,\Phi(\omega)}{d\omega^6} + \cdots .
\end{aligned}\right\} \tag{3}$$

Because of uncertainties of the higher order derivatives of the observed function it is customary, in practical applications, to neglect all terms following the second derivative in expression (3). Let us call the remaining formula the second order Eddington approximation.

It frequently happens, when the second order Eddington approximation is applied to 21 cm observations, that the resulting de-smoothed values of the hydrogen density (or of the optical depth) become negative, especially on the wings of the profiles.

There are several possible reasons why such negative values may occur:

1. Shortcomings of the mathematical technique of solving the integral equation.

2. Errors in the observed data.

3. Inadequacy of the assumed smoothing function to describe the smoothing effect.

In 21 cm observations one would expect all three of the above listed reasons to play a role. Their relative importance in affecting the de-smoothed results may be different for different observations, and depends for example on what region of the galaxy is observed. A tendency of the de-smoothing process is to enlarge fluctuations present in the observed curve. As a consequence errors in the observed data are enlarged, sometimes extensively, by the de-smoothing, besides the desired enlargement of the real variations in the observed curve. Correct de-smoothing of real variations without affecting the errors is indeed a difficult problem, which, in fact, has no unique solution as long as the relative contribution of the errors is not known. In chapter 3 of this paper we indicate a method by which this problem can be treated under the assumption of a gaussian smoothing function. This method can be generalized to any kind of smoothing function. A technique of de-smoothing of observed data by solving the integral equation exactly for simple local approximations to the observed curve is suggested in chapter 2.

2. Exact de-smoothing of local approximations to observed data

Let us approximate the observed curve in the region around $\omega = \omega_1$ by the function

$$\Phi(\omega) = A + B e^{-(\omega - \omega_1)/\eta}. \tag{4}$$

We determine the constants A, B and η from the observed value and from the first and second derivatives of the observed curve at the frequency ω_1. We then have

$$A = \Phi(\omega_1) - \frac{[\Phi'(\omega_1)]^2}{\Phi''(\omega_1)}, \tag{5}$$

$$B = \frac{[\Phi'(\omega_1)]^2}{\Phi''(\omega_1)}, \tag{6}$$

$$\eta = -\frac{\Phi'(\omega_1)}{\Phi''(\omega_1)}. \tag{7}$$

Because of the simple choice of the approximating function (4) we can then solve the integral equation (1) with the kernel (2) exactly. For the de-smoothed function we then obtain the following result:

$$\Psi(\omega_1) = \Phi(\omega_1) - \frac{[\Phi'(\omega_1)]^2}{\Phi''(\omega_1)} \left\{ 1 - e^{-\frac{1}{2}\sigma^2 \left[\frac{\Phi''(\omega_1)}{\Phi'(\omega_1)}\right]^3} \right\}. \tag{8}$$

The expression (8) should be of particular interest to apply to the wings of a 21 cm hydrogen profile, whereas it is not expected to give reliable results near a maximum of an observed curve, where relation (4) is not a natural approximation to choose. When the formula (8) is applied to an observed hydrogen profile

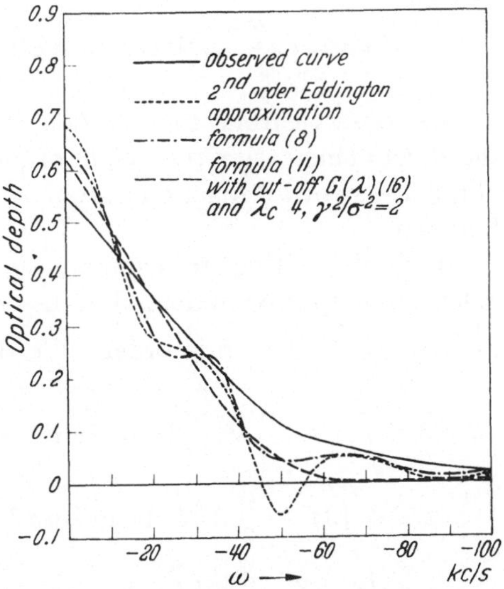

Fig. 1. Comparison between the results of the formula (8) and of the relation (11) assuming a smooth cutoff (16) ($\lambda_c = 4$ scale units of $5 \cdot 10^{-6}$ s, $\gamma^2/\sigma^2 = 2$, and $\sigma = 28.4$ kc/s) with the second order Eddington approximation. The observed hydrogen profile ($l = 122.5°$, $b = 2.5°$, old coordinates) was recorded at the Onsala Wave Propagation Observatory of Chalmers University of Technology

(see Fig. 1) it does not result in any negative values of the de-smoothed curve. Negative values appear, however, when the second order Eddington approximation is applied to the same curve.

Let us also remark that if the exponential function in relation (8) is expanded to first order in the exponent the result is identical with the Eddington approximation.

3. Fourier spectral method and the cut-off problem

The integral equation (1) with a gaussian kernel (2) has for $\Phi(\omega) = \cos \lambda\omega$ and $\Phi(\omega) = \sin \lambda\omega$ the solutions $\Psi(\omega) = e^{\frac{\sigma^2}{2}\lambda^2} \cos \lambda\omega$ and $\Psi(\omega) = e^{\frac{\sigma^2}{2}\lambda^2} \sin \lambda\omega$, respectively.

Let us represent an observed function $\Phi(\omega)$ by the Fourier integrals:

$$\Phi(\omega) = \int_0^\infty a(\lambda) \cos \lambda\omega \, d\lambda + \int_0^\infty b(\lambda) \sin \lambda\omega \, d\lambda \tag{9}$$

where

$$\left.\begin{aligned} a(\lambda) &= \frac{1}{\pi} \int_{-\infty}^\infty \Phi(\omega) \cos \lambda\omega \, d\omega, \\[2mm] b(\lambda) &= \frac{1}{\pi} \int_{-\infty}^\infty \Phi(\omega) \sin \lambda\omega \, d\omega. \end{aligned}\right\} \tag{10}$$

The formal solution of the integral equation (1) can then be written

$$\Psi(\omega) = \int_0^\infty e^{\frac{\sigma^2}{2}\lambda^2} a(\lambda) \cos \lambda\omega \, d\lambda + \int_0^\infty e^{\frac{\sigma^2}{2}\lambda^2} b(\lambda) \sin \lambda\omega \, d\lambda. \tag{11}$$

We observe from the relation (11) that the factor $e^{\frac{\sigma^2}{2}\lambda^2}$ has the effect of increasing the weight of the high-λ part of the spectrum of the de-smoothed function as compared with that of the smoothed function.

It is interesting to compare again with the second order Eddington approximation. Let us expand the exponential as $e^{\frac{\sigma^2}{2}\lambda^2} \approx 1 + \frac{\sigma^2}{2}\lambda^2$ to first order. We then have

$$\left.\begin{aligned} \Psi(\omega) &\approx \int_0^\infty \left(1 + \frac{\sigma^2}{2}\lambda^2\right) a(\lambda) \cos \lambda\omega \, d\lambda + \\ &\quad + \int_0^\infty \left(1 + \frac{\sigma^2}{2}\lambda^2\right) b(\lambda) \sin \lambda\omega \, d\lambda \\ &= \left(1 - \frac{\sigma^2}{2}\frac{d^2}{d\omega^2}\right)\left(\int_0^\infty a(\lambda) \cos \lambda\omega \, d\lambda + \int_0^\infty b(\lambda) \sin \lambda\omega \, d\lambda\right) \\ &= \Phi(\omega) - \frac{\sigma^2}{2}\frac{d^2 \Phi(\omega)}{d\omega^2} \end{aligned}\right\} \tag{12}$$

which is the second order Eddington approximation.

Let us now consider the relations (9) to (11) in somewhat more detail with regard to practical applications to observed data.

From the expression (11) we notice that $|a(\lambda)|$ and $|b(\lambda)|$ have to decrease at least as rapidly as $e^{-\frac{\sigma^2}{2}\lambda^2}$ for high values of λ in order for the integrals to be convergent. It follows that for values of λ above a certain limit where $e^{-\frac{\sigma^2}{2}\lambda^2}$ is less than a certain number, which we denote by ε, i.e.

$$\sigma\lambda > \sqrt{2\ln\varepsilon^{-1}} \tag{13}$$

the quantities $|a(\lambda)|$ and $|b(\lambda)|$ should be less than ε. We may regard the relation (13) as an uncertainty (or indeterminacy) relation. For a given value of ε, the domain of λ from zero to $\frac{1}{\sigma}\sqrt{2\ln\varepsilon^{-1}}$ contains more of the essential parts of the spectrum the larger the dispersion σ.

Accidental errors present in the observed curve may give relatively large contributions to the spectrum at high values of λ and cause the integrals of (11) to diverge or at least contain enlarged erors. It is therefore as a rule necessary to introduce a cut-off of the integration when the relation (11) is applied to 21 cm observations.

We may estimate a reasonable value at which to make the cut-off by requiring that the following two relations

$$\left.\begin{array}{c} \dfrac{a(\lambda)}{a_{\max}} < e^{-\frac{\sigma^2}{2}\lambda^2} \\[2em] \dfrac{b(\lambda)}{b_{\max}} < e^{-\frac{\sigma^2}{2}\lambda^2} \end{array}\right\} \tag{14}$$

and

are fulfilled.

Errors in the de-smoothed spectrum originating from accidental errors in the observed curve would be expected to play a dominating role for values of λ higher than the smallest of the two values

$$\left.\begin{array}{c} \lambda_a = \sqrt{2\ln\left(\dfrac{a_{\max}}{a(\lambda_a)}\right)} \\[2em] \lambda_b = \sqrt{2\ln\left(\dfrac{b_{\max}}{b(\lambda_b)}\right)}, \end{array}\right\} \tag{15}$$

and

which corresponds to the lower limit of λ in (14).

Numerical studies which we have done on the effect of a sharp cut-off when applied to the de-smoothed spectrum of an observed hydrogen profile indicate that such a cut-off is not advisable. The

120 H. Wilhelmsson:

reason is that because of the sharpness of the cut-off periodic oscillations are introduced in the de-smoothed curve.

To avoid this difficulty we instead introduce a smooth cut-off, defined by

$$G(\lambda) = e^{-\frac{\gamma^2}{2}(\lambda-\lambda_c)^2} + e^{-\frac{\sigma^2}{2}\lambda^2}\left[1 - e^{-\frac{\gamma^2}{2}(\lambda-\lambda_c)^2}\right] \tag{16}$$

and starting at $\lambda = \lambda_c$ in the integrals of relation (11). We choose the parameters γ and λ_c in such a manner that $G(\lambda)e^{\frac{\sigma^2}{2}\lambda^2}a(\lambda)$

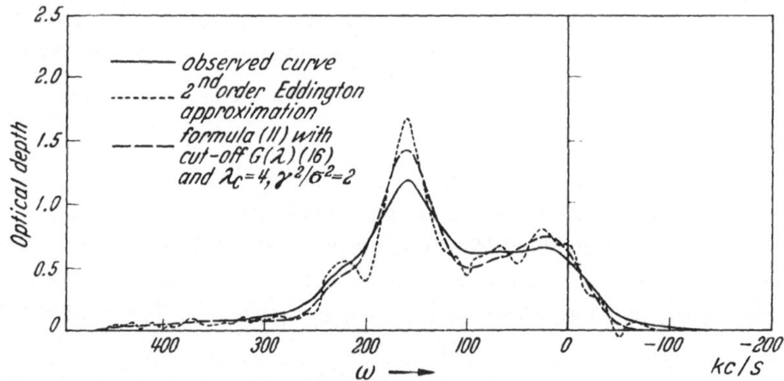

Fig. 2. Comparison between the result of relation (11) with smooth cut-off (16) and the second order Eddington approximation

and $G(\lambda)e^{\frac{\sigma^2}{2}\lambda^2}b(\lambda)$ are close to $a(\lambda)$ and $b(\lambda)$ (to which they join smoothly for increasing λ), respectively, at the value of λ defined by (15). By using this cut-off we leave but do not amplify the part of the spectrum associated with high values of λ.

In Fig. 2 we have plotted the result of the application of formula (11), assuming the smooth cut-off (16) with $\lambda_c = 4$ scale units (1 scale unit $= 5 \cdot 10^{-6}$ s), to an observed 21 cm profile ($l = 122,5°$, $b = 2,5°$, old coordinates) recorded at the Onsala Wave Propagation Observatory of Chalmers University of Technology. The result of the second order Eddington approximation, when applied to the same profile is also shown for comparison in Fig. 2. For the dispersion we have everywhere used the value $\sigma = 28.4$ kc/s. Part of Fig. 2 is plotted to enlarged scale in Fig. 1 together with the result of the formula 8.

Fig. 3 and 4 show how the spectral coefficients $a(\lambda)$ and $b(\lambda)$ of the observed curve plotted in Fig. 2 are modified by de-smoothing, assuming the smooth cut-off (16) with $\lambda_c = 4$ scale units and $\gamma^2/\sigma^2 = 2$. The de-smoothed spectral coefficients corresponding to the second order Eddington approximation are obtained by multiplying $a(\lambda)$ and $b(\lambda)$ by the factor $\left(1 + \frac{\sigma^2}{2}\lambda^2\right)$. We also show an $e^{-\frac{\sigma^2}{2}\lambda^2}$-dependence of interest in connection with relations (14)

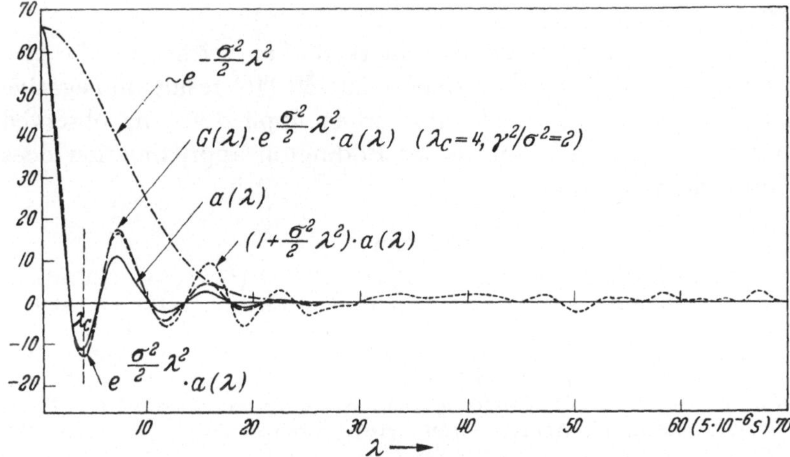

Fig. 3. The spectral function $a(\lambda)$ and its associated spectral function of the de-smoothed curve, assuming a smooth cut-off (16). The dotted curve is the spectrum of the second order Eddington approximation

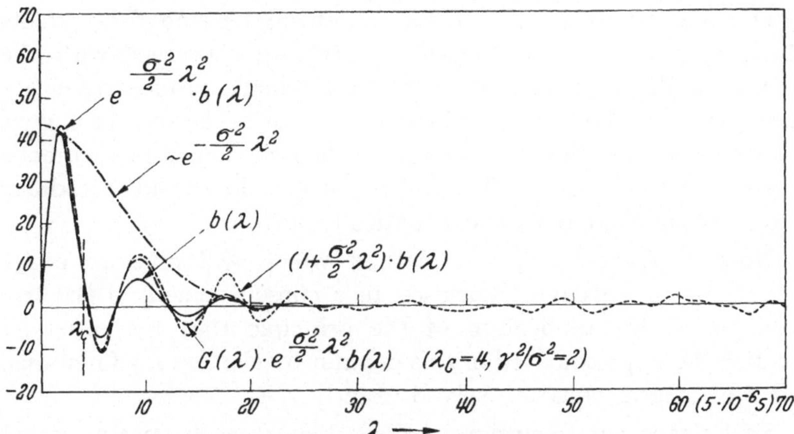

Fig. 4. The spectral function $b(\lambda)$ and its associated spectral function of the de-smoothed curve, assuming a smooth cut-off (16). The dotted curve is the spectrum of the second order Eddington approximation

and (15) and for the choice of cut-off parameters. Practically no amplification of the observed spectral coefficients occurs for larger than approximately 22 scale units with the smooth cut-off. The

fluctuations present in the spectrum of the Eddington approximation for $\lambda > 26$ scale units we consider as mainly enlarged errors
which originate from accidental fluctuations in the observed curve
of a kind that could not result from a smoothing of the assumed
type.

In conclusion it should be remarked that neither of the two
formulae (8) and (11) with a smooth cut-off (16) results in negative
values of the de-smoothed curve when applied to the observed
profile, whereas the second order Eddington approximation does
produce negative values.

References

[1] Schwarzschild, K.: Astron. Nachr. **185**, 81 (1910). — [2] Eddington, A. S.: Monthly Notices Roy. Astron. Soc. **73**, 359 (1913). — [3] Eddington, A. S.: Monthly Notices Roy. Astron. Soc. **100**, 354 (1940). —
[4] Malmqvist, K. G.: Arkiv **27**A, No. 24 (1941). — [5] van de Hulst, H. C.:
Bull. Astron. Inst. Neth. **10**, No. 367, 75 (1946). — [6] Ollongren, A., and
van de Hulst, H. C.: Bull. Astron. Inst. Neth. **13**, No. 475 196 (1957). —
[7] Wilhelmsson, H. Arkiv Astron., Roy. Swedish Acad. **3**, No 16, 187
(1963).

Discussion

Ollongren: To begin with I think Dr. Wilhelmsson should be
congratulated with the method for accounting for smoothing effects
which he has shown us to-day. Everybody concerned with the
process of de-smoothing of observed intensity profiles by either
Eddingtons method or an allied method using second or higher
differences knows what a nuisance the phenomenon of the occurence
of negative intensities is. They must be got rid of and procedures
used have been far from mathematically justified.

Now, however, we have a method which will give an exact
correction, no negative intensities, in a chosen demain of the fre-
quency-axis, by application of the principle that the observed
function be approximated in this domain by elementary functions,
and the integral equation solved exactly.

Still this is not a solution to all problems which arise in actual
practice. For: how are we to choose the domain for approximating
the known function, or, in other words, which criteria are to be
used to determine the domain or domains where the new procedure
can be applied to yield a better form of the function which we are
interested in? Could Dr. Wilhelmsson give some indication of
the way in which this can be done?

Wilhelmsson: The simple formula with an exponential term, which I gave as an example of the technique of approximating the observed function in such a way that the integral equation can be solved exactly, could preferably be used for the wings of the profiles, where the Eddington approximation gives negative densities. I have constructed similar formula more suited to represent the region around a maximum. The generalized spectral analysis, which I described, I think, forms a sound basis to discuss the influence of accidental errors. Such errors limit the region, over which the final integration should be extended, to a certain domain, the size of which depends also on the dispersion of the smoothing function.

D. Kuschbert (Tübingen): A Flying Spot Scanner for Star Counting and Determination of Star Distribution on Photographic Plates. (With 2 Figures.)

Zusammenfassung: Es wird eine ausführliche Übersicht über den Aufbau und die Arbeitsweise eines Lichtpunktabtastgeräts gegeben. Mit einem Lichtpunkt werden Sternphotoplatten Zeile für Zeile rasterförmig abgetastet. Die einzelnen Sternbilder liefern Zählsignale, die Aufschluß über Sternzahlen pro Flächeneinheit an der Sphäre geben. Die genaue Kenntnis dieser Zahlen bildet die Grundlage für stellarstatistische Rechnungen. Gegenüber der bisherigen Methode, die Sternbilder unter einem Mikroskop auszuzählen, bietet das rein elektronische Verfahren eines Lichtpunktabtasters den Vorteil großer Zählgeschwindigkeit bei mindestens ebenso großer Genauigkeit.

Up to date star counting and determination of star distribution is performed with the help of a microscope. According to the great number of stars these measurements take very much time. Therefore one was looking for a method to fasten these measuring processes by using electronic instruments, as we know that most investigations in Astronomy are done much quicker by help of such instruments.

One method, which appears most favourable, is the flying spot scanning. It is mainly used in television for projection of film exposures and works as follows: A very bright and intensive small light spot is scanning line by line over the image on the film and according to the blackening of the image the intensity of the light beam will be modulated. If one now projects this modulated light beam on the photocathode of a Multiplier Tube one will get voltage signals at the output resistance in respect to the modulation of the light beam, caused by the different amount of blackening of the image. The voltage signals are amplified and may be used for generation of an image on the screen of a television tube. In television the repetition frequency of the images is so quick that the human eye can not follow and one sees smooth movements if there are changes in the different image elements. This and the resolving power are the reasons, that one needs a horizontal light beam deflection frequency of about 15 kc/sec. According to the resolving power the bandwidth of amplifiers and magnetic stores has to be about 4 Mc. We know the difficulties to build magnetic tape recorders for storing television images. On the other hand they are very expensive.

As in our case — that means scanning of star photo plates — the velocity of deflection is not so important we lower the horizontal deflection frequency of our scanning screen drastically, but not too far, so that there is still a visible advantage against the method of counting star images on photo plates with a microscope. We choose a frequency of 100 c/sec. If our scanning screen consists

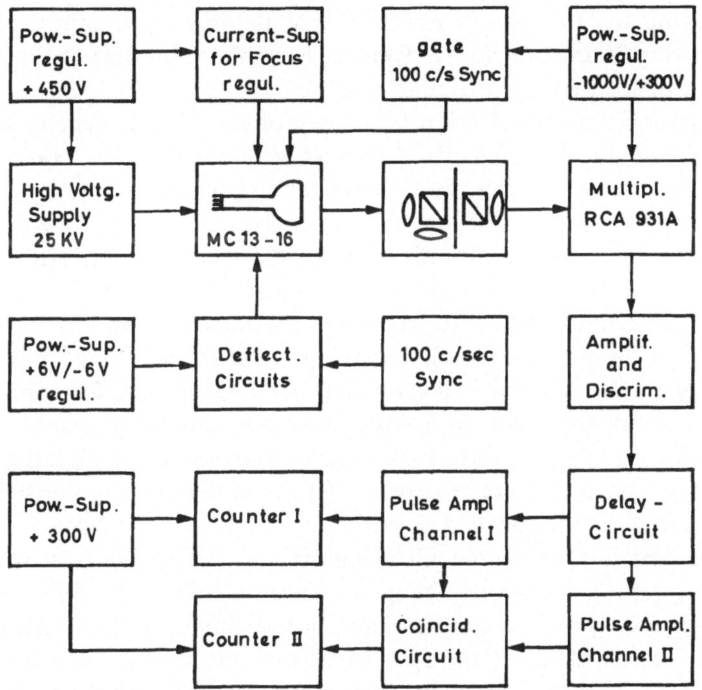

Fig. 1. Block Diagram representing the main features of a Flying Spot Scanner

of 1000 lines, the vertical deflection frequency will be 1/10 c/sec. Having a scanning light spot diameter of 10 microns and a screen of 1 × 1 cm the resulting bandwidth for amplifiers and magnetic stores is about 20 kc/sec, so that there arises no great difficulty in constructing devices. In a block diagram we shall explain the main features of a Flying Spot Scanner for star counting and determination of star distribution (Fig. 1).

Let us start with the Flying Spot Tube. It is a special development of a cathode ray tube on the screen of which one generates a small and bright light spot. The voltage supply is about 25 KV.

Focussing and deflection are performed magnetically as magnetic deflection and focussing give best linearity. To produce a constant intensity of the light spot over the whole tube screen it is necessary to regulate the high voltage supply. This is done by a feed back loop in the high voltage high frequency generator, which produces the voltage of 25 KV. The high voltage generator itself is fed by a regulated power supply of 450 Volts, which also feeds a circuit regulating the focussing current so that it remains constant. That is also very important for we want to have the same size of the light spot in every place of the tube screen.

The horizontal deflection frequency of 100 c/sec is synchronized by 50 c/sec line frequency. A ten stage dual devider, consisting of ten flip-flops devides the horizontal deflection frequency by a number of $100/_2 10$, so that we get a frequency of $1/10 \cdot 24$ c/sec, which is used for synchronization of the vertical deflection frequency. In that way we get a scanning screen of 1024 lines. The deflection circuits are transistorized. Because of the low vertical deflection frequency the deflection generator has to be coupled directly to the deflection coils which need about 1 A for complete deflection of the electron beam. A valve generator would then become very complicated. In the case of a transistorized deflection circuit we need a voltage supply of plus and minus 6 Volts, plus and minus for symmetric deflection.

Now having discussed all elements for generating a scanning screen on the cathode ray tube we want to follow the light beam through the optical system to the photocathode of the multiplier. On the screen of the flying spot tube the light spot scans a field of 8×8 cm. The spot diameter is about 80 μm. Through a lens we reduce the field to one cm², so that the light spot diameter will be 10 μm. Because of the resolving power of our scanner, we cannot make it smaller. The distance between the horizontal scanning lines will also be 10 μm. But before passing the photoplate the light beam is partly turned over by a prism to a multiplier of which the output signals are applied to the grid of the flying spot tube, thus regulating the intensity of the light spot in such a way that its intensity is always constant. Behind the photoplate a second prism splits the light beam so that one gets an image of the field to be measured on a ground glass disk for correct setting of the field on the plate. In the direct light path a field lens projects the aperture of the first projection lens on the photocathode of a multiplier, which turns the light signals modulated by the blackening of the star images to voltage signals.

The voltage signals at the output resistor of the photomultiplier are amplified and can be discriminated. Discrimination is necessary for determination of star distribution concerning their apparent magnitude. For star counting we have to apply the voltage signals to a delay line. We heard already that the spot diameter on the photo plate and the distance between the lines is 10 μm. Normally star images have diameters from 30 μm to 300 μm. That means that a single star image is going to generate several counting pulses instead of only one single pulse what we want. Therefore we have to compare two lines on the same place vertical to the horizontal deflection direction, to decide whether the signals coming from the two compared lines belong to the same star image. If they do so, only the first scan of one image has to be counted and all other signals from the same image have to be depressed.

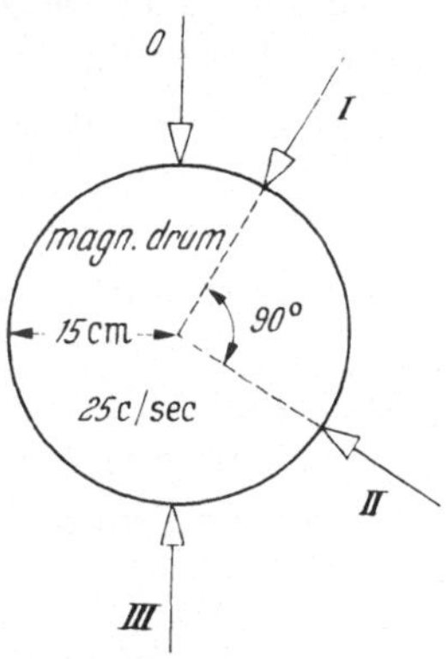

Fig. 2. Magnetic Drum for signal delay. 0 = writing coil, I, II and III reading coils

Our delay circuit is a magnetic drum (Fig. 2) rotating with 25 c/sec driven by a synchronmotor so that we have a good synchronization between drum rotating frequency and horizontal deflection frequency which was also synchronized by line frequency. In Fig. 2 we feed our signals over a coil 0 to the surface layer of the drum. Reading coil I will registrate one count over pulse amplifier I on counter I in channel I. If the drum has turned 90°, that means a time interval of 1/100 second or just the time for one scanning line, the signal will arrive at reading coil II. If the light spot hits still the same star image, there will be a signal at reading coil I. This signal at reading coil I will also be registrated on counter I but at the same time the coincidence of the signals from reading coil I and II will give one count on counter II. If we have scanned the whole field on the photo plate we only have to form the difference between counter I and counter II to get the correct number of stars in our field.

For determination of star distribution concerning their apparent magnitude we need the amplitude discriminator. We put it into a position that just the largest signals can pass that means that just the largest star images are counted. Now we enlarge its sensivity in steps which are equal to definite apparent magnitude values. The difference of two counts following one another will give us the number of stars of a definite magnitude interval. We only have to calibrate the discriminator with star images of known magnitude. Of course one might do the determination of star distribution with a multi-channel analyser but for our case such an instrument would become to expensive.

In this place we have still to say something about the resolving power of our flying spot scanner. The first point we have to regard was already mentioned formerly, when we explained the intensity regulation of the light spot. The graininess of the tube screen which might involve irregularities to the intensity of the light beam, are regulated in quite the same way. The second and more important point we have to remark is the graininess of the photographic plate. The diameter of the ligt spot was said to be 10 μm. If we now assume a photographic grain diameter of 1 μm the scattering $\sigma\,(S)$ of our results will be by a formula found by H. Siedentopf

$$\sigma(S) = \pm\, 0.7\, \frac{d}{D}\, \sqrt{S}$$

where D = spot diameter, d = diameter of grains, S = blackening. This is the reason why we cannot reduce the diameter to more than 10 μm, for then the signals generated by $\pm\,\sigma(S)$ would give counts too. On the other hand we only can resolve star image distances larger or equal 10 μm. Generally this assumption will not affect our measurements because one finds only very few star images touching or overlapping.

The instrument is still in development, but we made already so much experiences in the laboratory that we can say that star counting with a flying spot scanner brings much advantage against counting under a microscope. With some variances we also hope to do spectrum analysis of star spectras with the same instrument.

Discussion

Raimond: What happens if two star images are above each other so that they come at the same place at the drum? Is the second star counted in this case?

Kuschbert: This is a question of the resolving power, which is determined by the distance of the scanning lines and the spot diameter. Both are about 10 μm. Star images with distances of 10 μm or larger in any position, vertical or horizontal in regard to the scanned field, will be resolved. The resolving power might be increased by reducing the spot diameter and the line distances. But that is not desireable because then the graininess of the photographic plate generates larger noise signals so that one can not guarantee correct counting.

F. Potenza (Milano): MISA — A semiautomatic device for evaluation of spectra.

Summary: The paper deals with a system conceived by Prof. L. GRATTON and now under development at S.I.T. The subject deals with Mechanic details of the measuring stage, Optical layout, Digital readout of data, Operating sequence in measuring spectra.

Introduction

MISA, MIsuratore Semi Automatico (semiautomatic measuring device) is a system for fast, accurate measurement of line position on spectrum plates.

In large scale surveys of radial velocities of stars, plates can be grouped by spectral type, and radial velocities are then deduced from relative displacements of some selected lines from the neighboring in the reference spectra. In later types than A or F much time is spent in line identification, accurate centering, reading and writing coordinates. Subsequent work includes reducing abscissas to wavelengths in the reference and star spectra.

MISA operating sequence

To reduce time spent in such a survey we begun to build a spectrum comparator wich operates under the following operating sequence:

1. the first of many plates taken with the same spectrograph is set on the stage and the spectrum aligned with the stage direction of movement.

2. observing the image of the spectrum projected on the screen and explored by hand moving the stage, the operator stops on the line to be measured. Then, pushing a button, the approximate line position is punched on tape together with information on kind of spectrum (star, upper or lower reference).

3. item 2. is repeated on all interesting lines. Then tape is taken out from puncher and set in a reader. Now it shall operate like programmer. Any other plate of the series is set on the stage and aligned. The automatic coordinate reader is zeroed on the line chosen in 2. and the stage is brought to its start position. A push button order produces:

a) reading of the abscissa of the first line written on the tape and transferring the information to a buffer memory;

b) starting the stage in the increasing abscissa direction.

The stage moves until a coincidence circuit reveals identity between the number contained in the buffer memory and the one read in flight from the abscissa reading device. Coincidence stops fastly the stage. The line appears on the screen very probably uncentered because inertia of stage in stopping, doppler shift, ecc.

4. the operator centers the spectrum line on the axis of the projector by means of a photoelectric impersonal centering device described later. Then orders writing of the accurately determined coordinate, like in 2. If setting was unsatisfactory, another order can be given by which the coordinate punched on the tape is preceeded by an "error" code; when tape is read by a computer (or in MISA also) a block followed by the "error" code (then printed before the correct position) is neglected. The punched information can be contemporaneously typewritten.

5. item 4. is repeated until all the interesting lines of a plate are measured. Before writing information provaining from the subsequent plate, special codes can be hand introduced, if necessary by means of teletype to help the elaboration of data in the computer.

Construction

MISA is a stage with a memory of the position, automatic readout of the coordinate and with a photoelectric device for impersonal centering.

The stage. It is semikinematically supported, screw-driven; the screw is turned by hand during operation 1. and 2. When driven by tape program a motor actuates movements at a speed of 1 mm/sec. Spectra can be measured over a range of 100 mm. Longer spectra can be measured dividing them in parts no longer than 100 mm. Screw movements provide what is required for a fast centering of the plate. Spectrum is illuminated by a circular disc of polarized light and by a slit. A small portion of spectrum (ab. 8 mm diam.) is projected on a screen 20 times enlarged.

Impersonal centering. At the center of the screen the image of the illuminating slit on the spectrum falls in a device wich, with a fast response photometer, scans the intensity profile across the small portion of spectrum contained on the slit. The electrical output from the photometer is sent to the y of an oscilloscope whose sweep goes from left to right and from right to left alternately instead than always from left to right. When a line profile is slightly displaced from the center of this device, oscilloscope will show the line profile and its symmetrical not superimposed. Superposition can be obtained by hand moving the stage carrying the spectrum. Accuracy (depending also from line width) can be as good as $\frac{1}{2}$ micron or better, due to magnification of plate scale on the oscilloscope screen.

Automatic readout of the coordinate. A moiré fringes Ferranti grating system transforms movements of stage in pulse trains read by an up-down counter; accuracy is one micron. The number in the u.-d. counter (wich increases when moving stage to the right, otherwise decreases and can be zeroed everywhere) can be printed out by an order. The order of printing transfers the number accumulated in the u.-d. counter to a buffer memory. Each decimal character is serially taken out from the buffer memory, coded in the tape code and sent to the tape puncher. In a second, information concerning abscissa and kind of spectrum is written down. By means of a teletype (Flexowriter) connected to the puncher, other data can be added.

Position memory. The tape containing approximate positions of lines is read by a tape reader when operated by an order that starts contemporaneously the stage. The number is transferred to a buffer memory where is compared by means of a coincidence circuit to the number increasing in the u.-d. counter. Coincidence is reached first in the more significative character and after in all others. When all are coincident an "and" circuit stops motion in the stage.

Conclusion

The idea of speeding up wavelength measurements on spectrum plates was of L. Gratton[1]. He wrote down, together with A. Kranic[2] the specifications for the operating procedure of the device and sketched the optical and the electronic block scheme. MISA is now almost completed in Milano.

When adequate experience will be obtained in operating it, a digital photometer will be added to obtain transmission values at selected points of spectrum lines.

---

[1] Monte Mario and Monte Porzio Observatories, Roma, Italy.
[2] Osservatorio Astronomico Universitario, Bologna, Italy.

Discussion

Rohlfs: If the selection of the lines wich are to be measured is made by an electronic memory, this amounts up to the construction of a small computer. Would it not be cheaper to tranfer all the information on the intensity and position of spectral lines by punched tape to an allround computer and than later on do the selection and comparison of iron lines and stellar lines by the computer?

Potenza: In MISA, circuits comparable to the ones of a computer are at least ten times less (and costly). Tape out from a day of operation can be read and elaborated by a small computer (IBM 1620, Siemens, etc.) in less than an hour. The same computer attached to a simplified MISA sould be very badly exploited. Today punched tape can be converted in cards or magnetic tape by means of universal converters. Magnetic tape is too expensive an equipment to be directly attached to MISA. We think that punched tape is the best choice.

Høg: 1. How much does the electronic parts cost? 2. How long time does one measurement take?

Potenza: The electronics of MISA (so excluding optics, mechanics, tape reader and puncher) has a cost of about Lire (it.) 8 000 000 (about 13000 dollars). It includes: updown counter, coding and decoding, coincidence, buffer memories, serializers, power amplifiers and voltage supplies.

The measurement of a plate takes the following time (estimated):

plate setting and orienting	time (depends from observers ability)
1 finding next line	(1 — 5 sec) depending on mean frequency
2 adjusting position	(10 sec)
3 printing time	(1 sec)
4 dead times	(5 sec)
5 additional information (repeated measurement, missing line, measure to reject, and so on)	(10 sec)
2 + 3 + 4 gives	16 sec
1 mean (pessimistic)	5 sec
5 mean (pessimistic)	10 sec
total	31 sec

Høg: We have built in Hamburg a device for digitize and print the information from a microphotometer scan across 6 mm spectra taken with a prism objective. It costs much less than the complete MISA and will perform, sending its output to a computer, the function of fast recognizing spectral types.

Potenza: I think that your equipment is not comparable to MISA; it is a digitized microphotometer whilst MISA is a spectrum comparator; the prize you have declared means what you have spent to build it by yourselves and not what you should ask for selling. We are developing a digitized microphotometer wich will cost much more than yours, but will have a greater speed and capacity.

A. Gutsche (Offenbach): Die Verwendung der Lochkarte zur Speicherung und Auswertung von meteorologischen Beobachtungsreihen. (Mit 1 Textabbildung.)

Zusammenfassung: Nach einem kurzen Rückblick auf die Einführung der Lochkarte als Hilfsmittel für meteorologische Untersuchungen wird über Arbeitsmethoden und Erfahrungen der Lochkartenstelle im Zentralamt des Deutschen Wetterdienstes berichtet, die vorzugsweise für Aufgaben der Praxis tätig ist. Es werden Lochkartenschemata besprochen, die für das Ablochen von Beobachtungsreihen des klimatologischen, synoptischen und aerologischen Netzes verwendet werden. Ferner wird ein Überblick gegeben über die umfangreiche Lochkartensammlung des Zentralamtes und über die hier mit konventionellen Lochkartenmaschinen durchgeführten Auswertungsarbeiten.

Einige Jahrzehnte, nachdem der Deutschamerikaner HERMANN HOLLERITH auf den Gedanken gekommen war, Einzelangaben umfangreicher Kollektive einer Volkszählung nach einem bestimmten Schema durch Löcher in Karten darzustellen, fand dieses Verfahren auch Eingang in die Meteorologie [1]. Die Notwendigkeit für eine maschinelle Sortierung und Auswertung der meteorologischen Schiffsbeobachtungen, die zeitlich und räumlich stark zerstreut und darum einer Bearbeitung besonders schwer zugänglich sind, wurde bereits 1920 in England und wenige Jahre danach in den Niederlanden erkannt. Etwa 10 Jahre später begann auch die Deutsche Seewarte mit Vorbereitungen zur Lochung des sehr umfangreichen meteorologischen Beobachtungsmaterials von deutschen Schiffen [2].

Aber auch die Aufbereitung von Beobachtungsreihen der Landstationen für Praxis und Forschung der Meteorologie wurde früher vor allem dann zum Problem, wenn Häufigkeitsauszählungen gewünscht wurden. Es gelang POLLAK [3], [4] schon in der Mitte der zwanziger Jahre, im tschechosklowakischen Wetterdienst die laufende Lochung von meteorologischen Beobachtungen einzuführen. Zahlreiche Arbeiten POLLAKS aus dieser und etwas späterer Zeit basieren bereits auf Lochkartenauswertungen, wie z. B. seine Periodenanalyse der erdmagnetischen Charakterzahlen und seine Untersuchungen über die Häufigkeitsverteilungen des Luftdruckes an 111 europäischen Stationen.

Beim ehemaligen Reichsamt für Wetterdienst in Berlin wurde nach den ersten Versuchslochungen für flugklimatologische Zwecke im Jahre 1930 gegen Mitte der dreißiger Jahre in größerem Umfang mit der Ablochung synoptischer Beobachtungen von Landstationen begonnen. Damit konnten dringende Bedürfnisse der Flugklimatologie befriedigt werden, die besonders stark interessiert ist an Häufigkeitsverteilungen meteorologischer Elemente, welche zu den synoptischen Terminen im gesamten Stationsnetz gleichzeitig beobachtet werden. — Die folgenden Ausführungen basieren im wesentlichen auf den Arbeiten und Erfahrungen der Lochkartenstelle beim Zentralamt des Deutschen Wetterdienstes in Offenbach [5], [6]. Sie ist in erster Linie für die Aufgaben der angewandten Landklimatologie eingesetzt und hat in ihrem Archiv glücklicherweise noch einen großen Teil der früher in Berlin gelochten synoptischen Karten zur Verfügung.

Wenn man sich entschließt, ein Kollektiv von Beobachtungen auf Lochkarten zu übertragen, so tritt zunächst die Frage nach dem Lochkartenschema und einer eventuellen Anpassung der Lochunterlage an die Lochkarte auf. Während es bis vor kurzem üblich war, für die Lochkartenstelle des Seewetteramtes die meteorologischen Beobachtungsdaten der Schiffstagebücher zur Ablochung in besondere Tabellen zu übertragen, mußte beim Zentralamt auf eine derartige Hilfe ganz verzichtet werden. Das heißt, es mußte das Lochkartenschema immer möglichst weitgehend auf die abzulochenden Tabellen abgestimmt werden, die fast ausschließlich handgeschriebene Originale der Beobachter sind. Natürlich trägt diese Tatsache dazu bei, daß an das Lochpersonal relativ hohe Anforderungen gestellt werden müssen. Hinzu kommt eine erhebliche Erschwernis durch die Verwendung zahlreicher Steuerlöcher, beispielsweise zur Wiedergabe des Vorzeichens oder zur Erweiterung der Stellenanzahl eines Lochfeldes, sowie die Notwendigkeit, beim Lochen teilweise einfache Verschlüsselungen vorzunehmen.

Im allgemeinen wird angestrebt, zeitlich unmittelbar aufeinanderfolgende Daten vertikal, d. h. in verschiedene Lochkarten zu stanzen und nicht nebeneinander in eine Karte. Sofern für die Auswertung der Karten konventionelle Lochkartenmaschinen benutzt werden, bietet diese vertikale Lochung gegenüber der anderen Art, der sog. horizontalen Lochung, besondere Vorteile z. B. bei Häufigkeitsauszählungen. Um Lochkarten zu sparen oder um eine genügende Anpassung an die Lochunterlage zu erzielen, mußte leider mitunter für die horizontale Lochung entschieden werden. Das Bemühen, mit möglichst wenig Lochkarten auszukommen, kann sich aber leicht als falsche Sparsamkeit erweisen, weil dadurch die spätere Auswertung der Lochkarten oft erschwert wird.

Jede gelochte Karte wird geprüft, d. h. die manuelle Eintastung der Daten durch die Locherin am Motorlocher wird von der Prüferin am Motorprüfer wiederholt. Es gibt meteorologische Dienste, die bei der Datenübertragung auf Lochkarten eine andere Art der Prüfung vorziehen. Bis zu einem gewissen Grade kann man natürlich die Gültigkeit der Daten auch durch geeignete Arbeitsgänge in Auswertungsmaschinen prüfen. Von dieser vollmaschinellen Kontrolle kann jedoch nur erwartet werden, daß sie einen Teil der Fehler aufzeigt. Der versuchsweise vor einigen Jahren unternommene Verzicht auf die exakte Prüfung konnte nicht dazu ermutigen, die Kontrolle am Motorprüfer aufzugeben.

Im Deutschen Wetterdienst wird nur zentral in Hamburg und Offenbach gelocht, und zwar in 80spaltige Lochkarten. Das Mark-Sensing-Verfahren wurde nicht eingeführt. Bei dieser Methode werden zunächst die Daten manuell durch elektrisch leitfähige Striche mit Stift oder Tinte in eine Lochkarte übertragen, die anschließend maschinell in Löcher umgesetzt werden. Versuche ergaben, daß das Anstreichen der Lochfelder zu große Mühe erfordert, als daß es von den Beobachtern der meteorologischen Stationen verlangt werden könnte.

Im Zentralamt in Offenbach wird ein sehr großer Teil des laufend von den westdeutschen Stationen anfallenden Beobachtungsmaterials routinemäßig auf Lochkarten übertragen. Die Ablochung der beim Seewetteramt in Hamburg eingehenden meteorologischen Schiffstagebücher ist Aufgabe der dortigen Lochkartenstelle. Der Lochkartenbestand des Deutschen Wetterdienstes wächst insgesamt jährlich um mehr als $1^1/_2$ Millionen.

Das feste Lochprogramm der Offenbacher Lochkartenstelle umfaßt die Beobachtungen zu den drei Klimaterminen von annähernd 300 Stationen des Klimanetzes sowie die dreistündlichen bzw. stündlichen Beobachtungen von fast 100 Stationen des synoptischen Netzes. Laufend werden auch die aerologischen Beobachtungen, Registrierauswertungen und Luftdruckwerte an Koordinaten-Schnittpunkten der Nordhalbkugel gelocht. Insgesamt sind mehr als 30 Lochkartenschemata im Gebrauch.

Zu jedem synoptischen und zu jedem Klimatermin werden an den Stationen jeweils alle wichtigen meteorologischen Elemente

1348 / 4

Klima I

Card column groups (IBM punch card "Klima I"):

Luftdruck in Zehntel mm — Station — Jahr — Monat — Tag — Kartenart

Temperatur Extreme (Mittel, Max, Min., Tages-schwankung, Min. am Erdbod.)

Temperatur trocken in Zehntel Grad Celsius (I, II, III, Mittel)

Temperatur feucht (I, II, III)

Dampfdruck in Zehntel mm (I, II, III, Mittel)

Rel. Feuchte in % (I, II, III, Mittel)

Legend:

Luftdruck	600	:11	Steuerloch über Zehnerstelle
"	500	:11	" " Einerstelle
"	400	:11	" " Zehntelstelle
Temperatur < 0°		:11	" " Zehnerstelle
e bei Feuchttemp.		:11	" " Einerstelle
U = 100 %		:11	" " Zehntelstelle
unsichere Werte		:12	Überloch über Einerstelle

Monatsmittel — Lochung 44 in Spalte 2,3

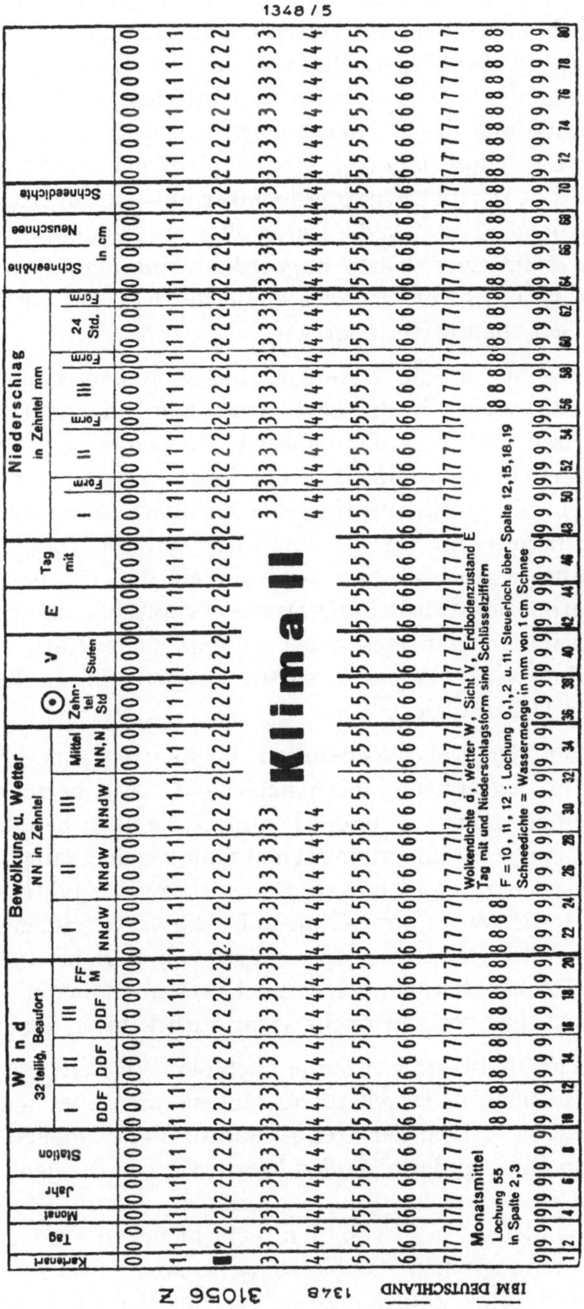

1348 / 5

Klima II

Wind 32 teilig, Beaufort: DDF I | DDF II | DDF III | FF M

Bewölkung u. Wetter NN in Zehntel: NNdW I | NNdW II | NNdW III | Mittel NN,N | ☉ Zehntel Std | V Stufen | E | Tag mit

Niederschlag in Zehntel mm: I Form | II Form | III Form | 24 Std. Form

Schneehöhe in cm | Neuschnee in cm | Schneedichte

Station | Jahr | Monat | Tag | Kartenart

Monatsmittel
Lochung 55 in Spalte 2,3

Wolkendichte d, Wetter W, Sicht V, Erdbodenzustand E
Tag mit und Niederschlagsform sind Schlüsselziffern
F = 10, 11, 12 : Lochung 0,1,2 u. 11. Steuerloch über Spalte 12,15,18,19
Schneedichte = Wassermenge in mm von 1 cm Schnee

IBM DEUTSCHLAND   1348   31056 Z

Abb. 1. Die Lochkarten Klima I und Klima II

beobachtet. Die synoptischen Meldungen lassen sich verhältnismäßig einfach lochen, da sie verschlüsselt, d. h. in sehr kompri-
mierter Form in den Tagebüchern vorliegen. Allerdings bedingt
eine Änderung des internationalen synoptischen Wetterschlüssels
auch eine Änderung des Lochkartenschemas, was hinsichtlich der
Lochkartenauswertung längerer Reihen nicht erwünscht ist. Für
die Beobachtungen eines synoptischen Termins ist auf einer Lochkarte ausreichend Platz. Da die Daten aller beobachteten Elemente
nebeneinander in einer Karte angeordnet sind, erhält man aus
Kollektiven synoptischer Lochkarten leicht Korrelationstabellen,
nach denen sehr häufig verlangt wird.

Mit Rücksicht auf die Datenanordnung in den Tabellen der
Klimastationen werden in die Lochkarten Klima I und II (Abbil-
dung) die drei täglichen Terminwerte der einzelnen Elemente
jeweils nebeneinander in eine Karte gelocht. Darum können aus
diesen Karten mit den konventionellen Lochkartenmaschinen Auszählungen nur getrennt für die einzelnen Termine vorgenommen
werden. Ferner ergibt sich hier der Nachteil, daß für die Aufstellung von Korrelationstabellen die Daten der miteinander zu korre-
lierenden Elemente erst maschinell in eine neue Kartenart zu-
sammengeführt werden müssen, sofern sie sich teils in Klima I,
teils in Klima II befinden.

Die Reihen der bisher genannten Lochkartenarten beginnen
größtenteils in den ersten Jahren nach 1945. Ein beträchtlicher
Teil der Reihen Klima I und II umfaßt jedoch bereits etwa
25 Jahre, und einige im klimatologischen Sinn wirklich lange Reihen
dieser Kartenart beginnen schon am Ende des vorigen Jahrhunderts.
So liegen z. B. die wegen ihrer Güte so häufig verwendeten Klima-
beobachtungen der Bamberger Sternwarte von 1891 bis 1958 voll-
ständig auf Lochkarten vor. Die Lochkartenreihe vom Hohen-
peißenberg im Alpenvorland reicht sogar zurück bis 1781.

Aerologische Meßwerte werden je nach dem Verwendungszweck
verlangt für festgelegte Druckstufen, Höhenstufen oder markante
Punkte. Es ist in keinem der drei genannten Fälle möglich, die in
der Vertikalen aufeinanderfolgenden Wertegruppen der gemessenen
Elemente auf einer Lochkarte unterzubringen. Darum werden für
einen Radiosondenaufstieg von den verschiedenen in Gebrauch
befindlichen aerologischen Lochkartenschemata jeweils mehrere
Karten benötigt. Ihre Zahl hängt ab von der Höhe des Aufstiegs

bzw. von der Zahl der markanten Punkte. Bei den Auswertungen dieser Karten treten oft schwierige Fragen auf; das ist z. B. dann der Fall, wenn Höhenwindmessungen niedriger Niveaus ausgewertet werden sollen in Abhängigkeit vom Vorhandensein der Meßwerte in einer größeren, nicht mehr regelmäßig erreichten Höhe.

Die Ablochung der Registrierauswertungen von westdeutschen meteorologischen Stationen umfaßt seit etwa 15 Jahren die Stundenwerte des Bodenwindes, der Temperatur und der relativen Feuchte. Neben dieser routinemäßigen Lochung werden in Sonderfällen auch stündliche Werte anderer registrierter Elemente auf Lochkarten übertragen. Im Hinblick auf die abzulochenden Tabellen konnten hierfür in der Regel nur Lochkartenschemata benutzt werden, in denen 12 oder 24 Stundenwerte eines Elementes nebeneinander angeordnet sind. Zur Untersuchung der Abhängigkeiten zwischen den Elementen ist es manchmal notwendig, aus den zunächst gewonnenen Einzelkollektiven durch maschinelle Doppelung ein neues Kartenkollektiv zu schaffen, in dem jede Karte die Daten verschiedener registrierter Elemente für eine bestimmte Stunde enthält.

Tägliche Luftdruckwerte an Koordinaten-Schnittpunkten liegen von einem fast die gesamte Nordhemisphäre umspannenden Netz seit der Jahrhundertwende, vom Raum zwischen Ural und Rocky Mountains sogar ab 1880 auf Lochkarten vor. Jede Karte enthält von einem bestimmten Tag und einem bestimmten Meridian die Daten an den Schnittpunkten mit den verschiedenen Breitenkreisen.

Im Vergleich hierzu stellen die seit dem Ende des vorigen Jahrhunderts vorliegenden Lochkarten mit den täglichen Kennziffern der Großwetterlage Mitteleuropas einen sehr kleinen Kartensatz dar. Weil jede Karte dieses Kollektivs nur die Kennziffer eines Tages enthält, können hieraus sehr leicht statistische Auswertungen gewonnen werden. Mitunter ist es notwendig, die Daten dieses Kartensatzes, der ebenfalls laufend wächst, zusätzlich in andere Lochkartenkollektive zu übertragen. Dadurch werden Auswertungen zur Untersuchung der Wetterlagenabhängigkeit einzelner Elemente ermöglicht. — In gleicher Weise wie die Kennziffern der Großwetterlage liegen übrigens auch erdmagnetische Charakterzahlen seit 1884 auf Lochkarten vor.

Spezielle Projekte erfordern gelegentlich die Lochung von umfangreichen Sonderkollektiven außerhalb des laufenden Programms. Zum Beispiel wurden für ein Gutachten über die Schneeverhältnisse des westdeutschen Straßennetzes die Schneebeobachtungen eines 20jährigen Zeitraums von 800 Stationen auf Lochkarten übertragen. In einer Karte sind jeweils tägliche Werte von sechs Stationen enthalten, die in der Schneebedeckung einander ähnlich sind. Tage ohne Schnee wurden bei der Lochung ausgeschlossen. Diese Einschränkung reduzierte zwar die Zahl der benötigten Karten, erschwerte jedoch deren Auswertung beträchtlich.

Für verschiedene Aufgaben der Hydrometeorologie wurden 5minutige Intensitäten des Niederschlags von einigen Stationen gelocht. Die Auswertung der Registrierstreifen konnte in diesem Fall ausnahmsweise besonders der Ablochung angepaßt werden. Eine Lochkarte dieses Kollektivs enthält jeweils für ein 5 Minutenintervall die Intensitätswerte von 16 Jahren. Für jedes Intervall des bearbeiteten Zeitraumes existiert ein Lochfeld, so daß sehr viele Null-Lochungen vorkommen. Diese direkte Erfassung der niederschlagsfreien Zeiten wird aber später die Auswertungsarbeiten erleichtern.

Im Offenbacher Lochkartenarchiv des Deutschen Wetterdienstes, das den Anforderungen entsprechend klimatisiert ist, lagern z. Zt. etwa 23 Millionen Karten. Das ständige Wachstum wird vielleicht dazu zwingen, äußerst selten benutzte Kollektive auf Mikrofilm aufzunehmen. Diesem Datenträger wird hinsichtlich seiner Beständigkeit gegenüber dem Magnetband beim amerikanischen Wetterdienst der Vorzug gegeben. Dort mußte man von dieser Möglichkeit der Raumersparnis bereits Gebrauch machen. Im Bedarfsfall können mit einem Spezialgerät aus dem Film wieder Lochkarten automatisch gestanzt werden.

Während der letzten Jahre gingen auch die meteorologischen Dienste junger Staaten dazu über, sich des Lochkartenverfahrens zu bedienen. Zum Beispiel konnte kürzlich auch in Ghana mit der Lochung begonnen werden. Die Meteorologische Weltorganisation empfiehlt kleineren Ländern mit dünnen Stationsnetzen eine Gemeinschaftsarbeit hinsichtlich der Lochkartenauswertung mit Nachbarstaaten und fördert auch den Lochkartenaustausch zwischen den einzelnen Diensten.

Die Bemühungen, internationale meteorologische Lochkarten festzulegen, sind bisher nur in einem Fall geglückt. In vielen Ländern sind nationale Schemata schon längere Zeit im Gebrauch. Die Homogenität einer Lochkartenreihe bestimmt deren Auswertungsmöglichkeiten in so hohem Maße, daß man sich zu Änderungen des Kartenschemas nur in dringenden Ausnahmefällen entschließen kann. Gegen die Umstellung auf internationale Lochkarten spricht ferner die Tatsache, daß in den einzelnen Staaten ganz unterschiedliche Anforderungen hinsichtlich der Lochkartenauswertung gestellt werden können. — Bisher wurde nur eine neue maritimmeteorologische Lochkarte international eingeführt. Das bedeutet jedoch nicht einen Verzicht auf die bisher benutzten nationalen Schemata für die Ablochung von Schiffsbeobachtungen. Die internationale Karte wird zusätzlich gelocht bzw. maschinell gedoppelt und ausgetauscht, so daß jede Nation, der auf internationalen Beschluß die Bearbeitung eines bestimmten Meeresgebietes zugewiesen ist, die Beobachtungen des betreffenden Raumes möglichst umfassend erhält.

Für die Lochkartenauswertung stehen dem Zentralamt in Offenbach eine Tabelliermaschine 421, ein Rechenlocher 602 und eine Statistikmaschine 101 der IBM mit den verschiedenen Zusatzmaschinen zur Verfügung. Zu letzteren gehören neben den Sortiermaschinen auch zwei Summenlocher für die Aufnahme von Ergebnissen der Tabellier- und Statistikmaschine in neue Lochkarten. Routineauswertungen, wie sie z. B. für das klimatologische Berichtswesen benötigt werden, spielen hier eine untergeordnete Rolle. Für die Drucklegung von Jahrbüchern wird im Ausland die Lochkarte teilweise in stärkerem Maße eingesetzt. Es ist natürlich möglich, Lochkartentabellierungen direkt als fertiges Manuskript zu benutzen. So leistet die Lochkartenstelle des Seewetteramtes mit ihren Tabellierungen eine wertvolle Hilfe für die jährliche Veröffentlichung der westdeutschen Feuerschiffsbeobachtungen. Bei der Drucklegung des Deutschen Meteorologischen Jahrbuches werden im ersten Teil Lochkartentabellierungen von Terminbeobachtungen ausgewählter Klimastationen für die Zusammenstellung des Manuskripts verwendet.

Auswertungsarbeiten werden in der Offenbacher Lochkartenstelle in erster Linie durchgeführt für die zahlreichen Auskünfte und Gutachten, welche Wirtschaft und Verkehr von den verschiedenen Zweigen der angewandten Klimatologie verlangen. Aber auch für viele andere Arbeitsgebiete des Deutschen Wetterdienstes werden Lochkartentabellierungen in großer Mannigfaltigkeit geliefert. Es überwiegen bei weitem Einzelarbeiten, für die das Arbeitsprogramm und damit auch die Schaltungen für die Auswertungsmaschinen jeweils neu entworfen werden müssen.

Summierungen der verschiedensten Art werden sehr häufig verlangt. Zumeist dienen sie der Mittelwertbildung. Neben den Summierungen für Zeitabschnitte fester Länge, z. B. für Tage, Pentaden, Jahreszeiten, Dezennien, werden auch Additionen für Zeitabschnitte variabler Länge benötigt, beispielsweise für zu warme und zu kalte Perioden oder für Zeiten, in denen bestimmte Strömungsverhältnisse herrschen. Gelegentlich sind aerologische Daten für Höhenschichten zu summieren und Luftdruckwerte an Koordinaten-Schnittpunkten oder andere Daten von einzelnen Stationen für bestimmte Areale der Erdoberfläche aufzuaddieren. Von wenigen Ausnahmen abgesehen, wird für derartige Arbeiten die Tabelliermaschine benutzt, die wie auch die Statistikmaschine die Ergebnisse anschreibt und, falls notwendig, von einem Summenlocher in neue Karten stanzen lassen kann.

Für Differenzbildungen wird die Tabelliermaschine ebenfalls häufig eingesetzt, daneben aber auch gern der Rechenlocher, weil er die Ergebnisse in die Ausgangskarte bzw. eine der beiden Ausgangskarten stanzen kann. In neuester Zeit wurden z. B. mehrmals Differenzen zwischen Werten von benachbarten Stationen in verschiedener Höhenlage und zwischen aerologischen Meßwerten aus verschiedenen Höhen zur Beurteilung der Stabilitätsverhältnisse in den unteren Luftschichten gebildet.

Häufigkeitsverteilungen können zwar auch mit der Tabelliermaschine gewonnen werden, in den meisten Fällen ist der Einsatz der Statistikmaschine jedoch vorzuziehen. Sie verarbeitet maximal 27000 Karten in der Stunde und ist mit 60 Zählern ausgerüstet. Sie vermag gleichzeitig Fälle mit bestimmten Merkmalen zu zählen und nach anderen Gesichtspunkten zu sortieren. Anschließend schreibt sie die Häufigkeiten an. In einem Kartendurchlauf kann sie mehrere Spalten berücksichtigen und Gruppen mit vorgegebenen mehrstelligen Merkmalsgrenzen auszählen und sortieren. Zum Beispiel liefert sie Auszählungen von Niederschlagsmengen nach einer logarithmischen Klasseneinteilung. Die Statistikmaschine wird besonders oft für die Tabellierung von Korrelationstabellen zur Untersuchung von Abhängigkeiten zwischen zwei Elementen benutzt. Hierbei ist allerdings eine Vorsortierung mit Einlegen von

Stoppkarten zwischen den Gruppen des einen Elementes notwendig. Andauerstatistiken werden dagegen fast nur mit der Tabelliermaschine aufgestellt.

Divisionen und Multiplikationen werden in der Regel mit dem Rechenlocher erledigt. Für Arbeiten wie die Bildung vertikaler Temperaturgradienten, die Komponentenzerlegung des Höhenwindes oder Divisionen für Mittelbildungen wird er fast ausschließlich benutzt.

Im allgemeinen ist es notwendig, für eine größere Lochkartenauswertung mit konventionellen Lochkartenmaschinen mehrere Maschinen möglichst sinnvoll neben- oder hintereinander einzusetzen.

Elektronische Rechenanlagen wurden für die Auswertung der Kollektive des Lochkartenarchivs in Offenbach bisher nur gelegentlich benutzt. Beispiele hierfür sind die Aufstellung von Mehrfachkorrelationstabellen für die Mittelfristvorhersage sowie eine Untersuchung von Luftdruckwellen der Nordhemisphäre mit dem Magnettrommelrechner IBM 650. Beide Arbeiten basieren auf den Lochkarten mit Luftdruckwerten an Koordinaten-Schnittpunkten.

Es ist geplant, in nicht sehr ferner Zukunft eine Großrechenanlage im Zentralamt des Deutschen Wetterdienstes aufzustellen. Diese Anlage soll in erster Linie zur Verarbeitung der synoptischen Wettermeldungen für die kurzfristige Prognose dienen und damit vor allem ermöglichen, den sprunghaft steigenden Bedürfnissen des Luftverkehrs gerecht zu werden. Neben der Verwirklichung des lang gehegten Wunsches von der versuchsweisen Erprobung zu einer routinemäßigen Anwendung von numerischen Methoden der Wettervorhersage überzugehen, darf vom Einsatz dieser Anlage im eigenen Haus auch eine Verbesserung von statistischen Methoden der mittelfristigen und wahrscheinlich auch der langfristigen Wettervorhersage erwartet werden. Auf zahlreichen Gebieten der meteorologischen Forschung und der praktischen Meteorologie können die Arbeitsmöglichkeiten durch die elektronische Datenverarbeitung erweitert werden.

Für die vielen relativ einfachen, in ihrer Art stark wechselnden Lochkartenarbeiten des Wetterdienstes wird jedoch auch bei Vorhandensein einer elektronischen Rechenanlage der Einsatz von konventionellen Auswertungsmaschinen weiterhin notwendig bleiben.

Literatur

[1] George, M. C.: An annotated bibliography of some early uses of punched cards in meteorology and climatology. Bull. Am. Meteorol. Soc. **26**, 76 (1945). — [2] Vring, E. v. d.: Das Lochkartenverfahren im Dienste der nautischen Meteorologie. Ann. Hydrogr. u. marit. Meteor. **60**, 115 (1932). — [3] Pollak, L. W.: Die Rationalisierung und Mechanisierung der Verwaltung und Verrechnung geophysikalischen Zahlenmaterials. Das Lochkartenverfahren. Naturwissenschaften **18**, 343 (1930). — [4] Pollak, L. W.: Further remarks on early uses of punched cards in meteorology and climatology. Bull. Am. Meteorol. Soc. **27**, 195 (1946). — [5] Guss, H., u. E. Reichel: Anwendungen des Lochkartenverfahrens in der Meteorologie. Ber. deut. Wetterdienstes US-Zone **2**, Nr. 12, 141 (1950). — [6] Guss, H.: Die meteorologischen Lochkarten des Deutschen Wetterdienstes und ihre hauptsächlichen Auswertungsmethoden. Meteorol. Abh. Freie Univ. Berlin **2**, H. 4, 2 (1955).

Siedentopf: We have now finished our official program, but I think we could spend still a little time for some questions connected with our problems that have not yet been discussed.

As first item I mention the evaluation of photographic plates. A semiautomatic iris-photometer with digital output can be seen here in the institute, similar devices have been realised elsewhere. Another problem is the recording of the density in photographic spectra to get the intensity distribution of the continuum or the intensity, the form and the position of spectral lines. I understand that Mr. Høg has done some work with digitizing a recording microphotometer at the Bergedorf Observatory, and so I ask him to tell us a bit about this instrument.

Høg: A microphotometer in Bergedorf has been equipped with digital output on punched cards. The equipment consists of an ordinary microphotometer (Krüss) with the plate driven by a

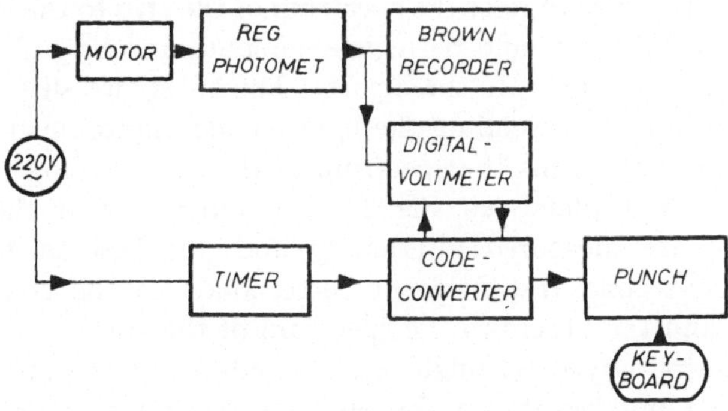

Fig. 1

synchronous motor and with output on a Brown Recorder. Parallel with this analog output runs a digital output built up of a Beckman Digital Volt-Ohm Meter (Model 5350 H), a code converter and an IBM card punch (Model 024 with input-output device), see Fig. 1. We can for instance take 2.4 measurements per second corresponding to 1 measurement per 0.015 mm on the plate. This means that a spectrum of 580 A/mm taken with the big Schmidt with objective prism can be registered in less than 3 minutes and gives about

20 punched cards with about 500 measurements of 3 decimal digits each. After the registration of each spectrum we shall punch on the next card the number of the star and the zero-point of the wavelengths. These cards containing the total information of the spectra can be fed into the IBM 650 where a program, not yet finished shall derive a two-dimensional classification of the stars using the criteria developped by LINDBLAD and others.

This digital output has cost 8000 DM and additionally 3000 DM a year for the IBM punch. It has been much used to digitize the iris-reading of the Sartorius photometer.

The measurement of radial velocities by slit spectra could be accomplished by a microphotometer with an accurate movement

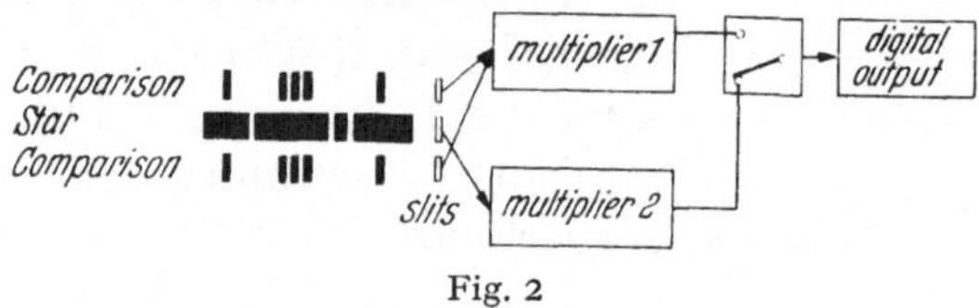

Fig. 2

of the plate and a digital output similar to that described above. But here a very accurate timing and a higher speed of output is necessary. In order to refer the spectrum of the star to the comparison spectrum these should be registered simultaneously. We could use a long slit divided into three parts: The outer two slits scan the comparison spectrum sending the light to one photomultiplier, the slit in the middle scans the spectrum of the star sending the light to another multiplier, see Fig. 2. The output from these two multipliers are measured alternately and punched on tape, on which each second measurement corresponds to the comparison spectrum and the others to the spectrum of the star.

For digital output we might use the pulse-counting photometer described in my lecture on the photoelectric micrometer for the Bergedorf meridian circle. Its price of 50000 DM is not prohibitive since it is used at night at the meridian circle. Its speed of 10 measurements per second would permit the registration of a 20 mm slit spectrum in 7 minutes giving 1 measurement per 0.01 mm in the comparison as well as in the star spectrum. A high speed computer could in a purely impersonal way derive the radial velocity as well as a classification of the star.

These examples indicate that we should consider the use of simple digital scanning equipment and of the great flexibility of digital computers by the measurement of spectra. This might in many cases for reasons of accuracy and cost be preferable to mechanically sophisticated equipment like spectrocomparators or machines with photoelectric setting on lines.

Siedentopf: With digital voltmeters as well as with voltage frequency converters the number of possible observations per second can be made as great as the electromagnetic device for punching cards of tape allows. With our Lorenz Schnellocher SL 614 which does up to 40 punches a second only 5 observations per second can be recorded since one observation with the necessary commands for giving the data from the tape to a teletyper needs 8 punches. There are other tape punchers on the market, e.g. Facit, which permit up to 150 punches a second and therefore about 20 observations per second. Quick digital voltmeters give 200 or more readings per second, to store this high information rate a magnetic tape recorder is necessary. If we scan a photographic spectrum with a resolution of 20 microns or 50 readings per mm the time needed for scanning a spectrum of 100 mm length is only about 25 seconds.

Potenza: I think that such a device is not useful. All astronomical telescopes of the world cannot produce more than one spectrum per minute.

Siedentopf: But we must not forget that a single exposure with a Schmidt camera and objective prism gives already many thousands of spectra.

We will now discuss the handling of time series with rapid variations of the observed function as they occur in atmospheric microturbulence, seeing and scintillation. This problem is closely related to the scanning of photographic densities, but whereas we can take our own time for the scanning of a photographic spectrum, we find in the case of turbulence effects that the phenomena themselves determine the necessary number of observations per unit time. If we wish to digitize the observations we have to remember the sampling theorem: if a function $g(t)$ does not contain any frequencies greater than W c/s in the Fourier transformation it is completely determined by the values of the ordinates in a distance $\frac{1}{2}$ W sec. Stellar scintillation e.g. contains frequencies up to about 300 c/s, so 600 observations of the momentary brightness per second are necessary. The number of discernable values for the ordinate is given by the signal to noise ratio S/N of the observations and for the resulting flux of information H we find

$$H = W \operatorname{ld}\left(1 + \frac{S}{N}\right) \text{ bits/sec}.$$

In the case of stellar scintillation we get for a S/N ratio of about 30 a value $H = 1500$ bits/sec, an information rate that can easily be handed by analogous recording but needs an expansive quick digital voltmeter with converter for magnetic tape in the case of digital recording. This tape can be fed to an electronic computer to obtain e.g. the frequency spectrum of the scintillation, the autocorrelation function or any other characteristic parameters of the time function.

In general the values of the time function are of less interest than the corresponding frequency spectrum. So an immediate analogue record of the frequency spectrum by a wave form analyser is often of advantage. If only the distribution of the values of the oscillation function is of interest a resolution of the function into aequidistant pulses, $\frac{1}{2} W$ sec apart, and the use of a pulse height discriminator with counting the pulses of different heights give an immediate result of good accuracy. These examples show that in each single case of observations it has carefully to be considered whether an analogue method or the digitizing of the observed values should be prefered.

We will now touch another question, the measurement of distances by automatic electronic methods. This can be done either on a photographic plate or directly on the sky, e.g. observations of double stars, of parallaxes or of the distances of any celestial body from a reference object. For the determination of positions on photographic plates measuring engines with printed digital output are already commercially available. The flying spot scanner described by Mr. KUSCHBERT also gives principally a possibility of observing the differences in coordinates of two objects very quickly but with limited accuracy. If the scanning period is 10 seconds for 1000 lines and we measure the time difference between the passage of the flying spot over the two objects in units of 10^{-5} sec we get a 6 digit reading. The first three digits give the x, the last three digits the y when the scanning is parallel to the x-axis.

The use of a similar device is being prepared for our observations of seeing. By a double telescope two images of a star are produced on a film moving in steps of 1 mm, the movement being connected with the shutter. The exposure time is 1/10 or 1/25 of a second, the time between two exposures 1 second. The fluctuation of the distances between the two images determine the value of the seeing parameter. To observe the distances the images are projected on a slit through a slowly rotating quadratic prism and the time intervall between the two passages recorded by an electronic counter. There are different methods to get mean square value or the distribution function of the distances but I cannot enter into these details.

I only should like to say that the methods for recording distances on photographic plates cannot be used for stars at the sky. Direct photoelectric distance observations need at any rate some integrating device since it is not possible to get a sufficiently accurate value for the position in less than about 30 seconds. It is therefore necessary to use similar methods as have been mentioned in our discussion of the photoelectric meridian circle.

Schwarz: I hear very often the problem of digitation is the problem to get a distribution function of some varying parameters, and it is very useful sometimes to get the exact order of the events. For instance here with this problem you will have distances of two pulses. Then it could be used an instrument which is well-known in nuclear physics, it is called a pulse height analyser which selects pulses of different amplitude and gives a distribution function of the pulse height. In this special case you mean, you have to transform the height distance in an amplitude voltage — in a pulse of a well determined amplitude — and then you give these pulses of different amplitude to a pulse height analyser. Then you get the exact distribution of the occurrence of different amplitudes of the pulses which means that you have a distribution of the time interval between these two pulses. Another example which was mentioned today this was the flying spotrecorder of photoplates. You get different pulses of different height; you can use such a pulse height analyser too to record the amplitude of these pulses. But it is not necessary to know exactly when occurs the smallest or the biggest, it doesn't matter. There you have the integral of the distribution in which you are interested. I don't know if in astronomy this machine is known and used but I think it could be very useful.

Siedentopf: With the first pulse you start a rising voltage and you stop this rise with a second pulse. Then you have in this point a certain voltage. You use this voltage for the pulse height analysis. Yes, this principle is known. But it is also a question of the price. A pulse height analyser is a very expensive instrument.

Tucker: I wonder if Prof. SIEDENTOPF can tell us any news of the experiments made in the Soviet Union on this problem of measuring linear intervals. There is a report concerning a device for measuring the linear intervals between two divisions. The image of the two divisions is moved uniformly with the help of a optical-mechanical device relative to the slit. When the image attains the site of the slit, it forms an impuls, special generated signals account it. The next of the measurement interval is 0.01. And they suggested if they wanted more accuracy to do it again and take the mean. I am thinking of applying it to the automatic registration of scales and graduated circles. Also in an application to time we are interested in direct reading of the graduation of meridian circles. I was disappointed to find that there were no delegates here from the Soviet Union so that I could have asked them. But because Prof. SIEDENTOPF has mentioned that he is working on the same field I wonder if he is having news of what is being done in Pulkovo.

Siedentopf: No, I am sorry I have no news about it.

Siedentopf: We are now through with our program and our general discussion. It remains for me only to express my and our all thanks to those who have contributed to this colloquium, especially to those who have acted as chairman in the four sessions and who have presented papers.

It is of course somewhat difficult to judge the result of our meeting but I hope that you will return from Tübingen with the feeling that our experiment has not been without success. It is a general experience that the contact of ideas of different persons gives rise to a next generation of ideas, and so I hope that you leave here with some new ideas in your mind, ideas that may bring further progress to automatic devices for astronomical instruments and so to our science in general.

Liste der Teilnehmer am Symposium

ADAMS, A. N., Washington
BACCHUS, P., Strasbourg
BAHNER, K., Heidelberg
BARANNE, A., Marseille
Baumann, W., München
BEHR, A., Göttingen
BONNIN, J.-CL., Paris
BRUZEK, A., Freiburg
ELVIUS, T., Uppsala
ERIKSSON, P.-J., Uppsala
FENKART, R., Basel
FRICKE, W., Heidelberg
GIESE, R.-H., Tübingen
GIRNSTEIN, H.-G., Bonn
GONDOLATSCH, F., Heidelberg
DE GRAAFF, W., Utrecht
DE GROOT, T., Utrecht
GUTSCHE, A., Offenbach a. M.
GUINOT, B., Paris
HANSSON, N., Lund
HUBENET, H., Utrecht
HØG, E., Hamburg
KARLSSON, B., Lund
KÖHLER, H., Oberkochen
KRÄMER, G., Tübingen
KÜHNE, C., Berlin-Mariendorf
KUSCHBERT, D., Tübingen
LACROUTE, P., Strasbourg
LINDBLAD, P. O., Saltsjöbaden
    (Stockholm)
VON LUDWIGER, J., Hildesheim

MATTIG, W., Freiburg
MAYER, U., Tübingen
OLLENGREN, A., Leiden
PAPERLEIN, D., Tübingen
PLESSE, H., Oberkochen
POTENZA, F., Rom
PRÉVOT, L., Marseille
RAIMOND, E., Leiden
RAKOSCH, K., Graz
REYNEN, J., Brussels
ROHLFS, K., Bonn
SCHEFFLER, H., Tübingen
SCHNEDLER-NIELSEN, Broerfelde
    (Dänemark)
SCHMIDT-KALER, TH., Bonn
SCHUBART, J., Heidelberg
SCHULER, W., Neuchâtel
SCHWARZ, U., Paterswolde (Holland)
SCHWESINGER, G., Oberkochen
SIEDENTOPF, H., Tübingen
SINNERSTAD, U., Saltsjöbaden
    (Stockholm)
STEGLICH, K., Berlin-Mariendorf
STRAND, K. AA., Washington
TAMMANN, G. A., Basel
TUCKER, R. H., Herstmonceux
    (England)
WALTER, H., München
WELIACHEW, L., Mendon
WERNER, H., Oberkochen
WILHELMSSON, H., Göteborg